'This book makes some of Chris Smout's finest scholarship available to the wider readership it richly deserves. These collected writings mark a major advance for the cause of environmental history in the UK and embody what the author memorably referred to in a previous book, *Nature Contested*, as "available history" whose purpose is "neither to justify nor direct the present, but to inform it".'
Professor Peter Coates, University of Bristol

'The British Isles are one of the world's most beautiful landscapes, and Chris Smout is a wonderful guide to their history of humans interacting with nature. Brilliantly written, lovingly told, and rigorously imagined, these essays are a revelation.'
Donald Worster, author of *A Passion for Nature: The Life of John Muir*

'That the book should end with provoking thought in its readers cannot be anything but good and reinforces our sense of gratitude that these essays should exist and that the publishers should bring them all together.'
Ian Simmons, *Environment and History*

'Good, scholarly and accessible environmental history like this does carry with it empowerment, vision and the very necessary context (that Nature, like us, has a history) that can only benefit those charged in society and politics with planning our future. I will certainly be encouraging my under-graduate and postgraduate environmental history students to dip into this book: they are after all (whether they welcome it or not!) the next environ-mental generation.'
Rod Lambert, *Landscape History*

'The modest title of this book gives little idea of the excitements that lie within ... This is a memorable book, rich in scholarship and full of argument, and elegantly written ... the brilliance of the essays must make *Exploring Environmental History* a thoroughly worthwhile purchase or gift.'
Paul Ramsey, *Recorder News*

'*Exploring Environmental History* is a collection of essays and papers which distil the professor's latter-day researches and reflections. They are charac-teristically acute and uncompromising.'
Roger Hutchinson, *Scottish Review of Books*

D1353449

T. C. Smout is Emeritus Professor of Scottish History at the University of St Andrews and Historiographer Royal for Scotland. He became an environmental historian in the 1980s and in 1999 was invited to deliver the Ford Lectures in Oxford, published later by Edinburgh University Press as *Nature Contested: Environmental History in Scotland and Northern England since 1600*. He has also written extensively on woodland history, most recently *A History of the Native Woodlands of Scotland, 1500–1920* with Alan R. MacDonald and Fiona Watson (Edinburgh University Press, 2005). His earlier works on Scottish social history, *The History of the Scottish People, 1560–1830* (1969) and *A Century of the Scottish People, 1830–1950* (1986), are regarded as classics.

Exploring Environmental History

Selected Essays

T. C. Smout

EDINBURGH UNIVERSITY PRESS

For Margaret
For many millions of words

© T. C. Smout, 2009, 2011

First published in 2009 by
Edinburgh University Press Ltd
22 George Square, Edinburgh
www.euppublishing.com

This paperback edition 2011

Typeset in Minion and Gill Sans
by Servis Filmsetting Ltd, Stockport, Cheshire and
printed and bound in Great Britain
by CPI Antony Rowe, Chippenham and Eastbourne

A CIP record for this book is available from the British Library

ISBN 978 0 7486 4561 9 (paperback)

Contents

Acknowledgements

In writing these essays over a long time period, I have incurred more debts to friends and fellow scholars than I can name, but I would particularly like to acknowledge the help and support of the following: for Chapter 2, John Sheail, David Jenkins, and the late John Berry and Morton Boyd; Chapter 3, Fiona Watson and Oliver Rackham; Chapter 4, Richard Tipping and the anonymous referee of the original article; Chapter 5, Douglas Malcolm; Chapter 6, Hugh Ingram and Alexander Fenton; Chapter 7, David Dickson, Cormac O'Grada, Ole Lindquist and the late Magnus Magnusson; Chapter 10, Peter Coates; Chapter 11, Guy Anderson; and Chapter 12, Marcus Hall and Frans Vera. For those at EUP and for everyone else who helped, great thanks as well. To Margaret Richards who typed and managed so many drafts and revisions over the years, the biggest thanks of all.

For acknowledgements to publishers and others who generously allowed the reprinting and revision of the originals, and also of the photographic illustrations, please see the chapters individually.

Illustrations

Introduction

The essays collected here were written over almost two decades, originating generally as conference or seminar presentations and subsequently enlarged for publication, though one, the longest, was the British Academy Raleigh Lecture for 1993, another was a contribution to a festschrift to Louis Cullen in Dublin and the last was specially written for a volume in memory of John Morton Boyd, one time head of the Nature Conservancy Council in Scotland. They focus on Britain in the five centuries since 1600, and especially on Scotland, though in one case they embrace Ireland and Iceland as well. Their themes reflect the manner in which I came to environmental history after retirement from a working life as a Scottish social and economic historian. Several are about how people have used natural resources and with what consequence for the resource: soils, bogs and especially woods. I owe the last focus partly to the inspiration of reading Oliver Rackham and partly to my own earlier interests as a historian of trade and of rural life. Others deal with the interactions between people and other species, a topic to which I was drawn by my love of birds and by my involvement in the 1980s and 1990s with the Nature Conservancy Council and its successor, Scottish Natural Heritage. I learned there in boards and committees how deep and complex were the issues surrounding nature conservation, and how little it was appreciated that they were matters of history as well as of contemporary manoeuvring and posturing round vested interest. One essay, the first, is bibliographical, trying to explore the history and character of environmental history within British scholarship. All the essays have been revised for publication in this volume, some very extensively, partly to take account of new evidence and partly to minimise overlap between them.

Environmental history is an extremely broad, even amorphous, subject. Some have said that it is not so much a subject in its own right as a theatre where other subjects appear and interact, and certainly the most exciting environmental history is often interdisciplinary. Environmental history is concerned about the relationship between people and environment, but unlike economic history or historical geography (to which it is closely related), the environment itself shares centre stage with the people who use it. It is the history of wolf hunting from the perspective of what happens to the wolf as much as what happens to the hunter. It is the history of the climate when sent off course by sunspots and volcanoes as well as by carbon emissions, but also the history of human response to climate change. The history of waste, of the transition from a world of recycling to a world of rubbish disposal, could be packaged as urban history, but not if you consider how the implications go from the town to the wider environment. It has a great deal of overlap with other subjects, but an energy of its own and a purpose. This is to place man in the context of his environment, not as a master with dominion over nature but as part of nature and subject to its laws, and further to show how his actions have impacted on nature.

Environmental historians, the youngest siblings among Clio's children, feel insecure and are much given to introspection. They speculate, often with a high degree of sophistication,[1] about what their subject is and where it should go next, and why 'mainstream' history does not afford it the consideration they consider it deserves. Sörlin and Warde argue for deeper engagement with social and political theory.[2] Mei Xueqin draws a distinction between three kinds of research on the environment: the purely scientific, which she terms natural history; the history of the environment as resource use from the perspective of the social sciences, such as geography and economic history; and environmental history proper, defined as the study of the relationship between humanity and nature. It is, she says, inescapably anthropocentric because that is what history is, and she is probably right.[3] Yet methinks we all worry too much, and we should accept environmental history in all its variety and confusion, embracing the sciences and the social sciences, making it an interactive performance. That is where the excitement lies.

Let us accept, then, that environmental history is closely related to economic history, cultural history and historical geography, but distinguished by its focus, and also that these distinctions can be hard to perceive. Any examination of evolving attitudes to nature, for instance, is cultural history, but since our attitudes have become so fundamental to the fate of the non-human world, it is also environmental history. Like other big subjects – say, social history – it has no single unifying theme, but many aspects that can be developed, often with help from others. Take the history of sustainability, one of the most neglected aspects of our subject but one urgently in need of development. Since it is the history of resource use viewed from a certain perspective, it calls for the input of the geographer and economic historian, and

---

[1] See 'The nature of environmental history', *History and Theory*, Theme Issue, 42 (2003).
[2] S. Sörlin and P. Warde, 'The problem of the problem of environmental history: a re-reading of the field', *Environmental History*, 12 (2007), pp. 107–30.
[3] Mei Xueqin, 'From the history of the environment to environmental history: a personal understanding of environmental history', *Frontiers of History in China*, 2 (2007), pp. 121–44.

indeed it would be difficult to pursue it without regard to the work of Tony Wrigley (both geographer and economic historian) and what he has termed the transition from an organic to an inorganic economy, in Britain spanning the period from the sixteenth to the nineteenth centuries.[4] The work of Paul Warde on British energy history exemplifies such an approach.[5]

Crucial to the advancement of environmental history in many of its aspects, is engagement and collaboration with science. The Americans, and to a much smaller degree the British, distinguish between ecological history (largely pursued by scientists) and environmental history (pursued by humanists). This unhelpfully encourages each to stay either side of the science/arts divide instead of collaborating to use one another's methodologies and sources.[6] For instance, palynology, the study of fossil pollens, will tell you things about the history of a wood no document can ever reveal, like its changing species composition or ancient episodes of clearfell. But documents can demonstrate what the palynologist can only speculate about: why, in eighteenth-century Argyll, oak pollen declines – the oak did not decline but was being coppiced on a rotation too short to allow for flowering.[7] There are not many polymaths who can take both science and history on board in their own skill set (there are some, like Oliver Rackham). But it is not intellectually difficult to set up collaboration between scientists and historians on a research topic,[8] though it may be institutionally difficult. For all the good words about encouraging interdisciplinary research, when push comes to shove, schools, faculties and universities are often unwilling to cross the zealously defended boundaries, or research councils to award money for work that lies partly outwith a narrow remit. If they do, the institutions may next fall out quarrelling over the distribution of the money. The social anthropology of academe is troublesome, but such difficulties should not deter us from trying.

---

[4] E. A. Wrigley, *Continuity, Chance and Change: the Character of the Industrial Revolution in England* (Cambridge, 1988); E. A. Wrigley, 'Meeting human energy needs: constraints, opportunities and effects', in P. Slack (ed.), *Environments and Historical Change: the Linacre Lectures, 1998* (Oxford, 1999), pp. 76–95; E. A. Wrigley, 'The transition to an advanced organic economy: half a millennium of English agriculture', *Economic History Review*, 59 (2006), pp. 435–80.

[5] P. Warde, *Energy Consumption in England and Wales, 1560–2000* (Naples, 2008).

[6] See for example, D. Egan and E. A. Howell, *The Historical Ecology Handbook* (Washington, DC, 2005).

[7] P. Sansum, 'Argyll oakwoods: use and ecological change, 1000 to 2000 AD – a palynological-historical investigation', in *Botanical Journal of Scotland*, Atlantic Oakwoods symposium, 57 (2005), pp. 83–98.

[8] For a successful example, see A. L. Davies and P. Dixon, 'Reading the pastoral landscape: palynological and historical evidence for the impacts of long-term grazing on Wether Hill, Ingram, Northumberland', *Landscape History*, 29 (2007), pp. 35–45.

Environmental history is often understood as if it first emerged in America in the 1970s, and although this is something of an oversimplification, the American initiative, with its enormous talent and crusading energy, set its stamp on the worldwide development of the subject for at least a generation. Its first driving theme was the impact of man on wilderness, so appropriate to the history of the United States, and readily translated to the histories of Australia (where it had already begun to be independently explored),[9] and to Africa and India where it could be conveniently spliced with imperial history. It was less appropriate to the study of long-settled and less overwhelmed countries like Europe, China and Japan, but they in due course developed their own ways of looking at their environmental histories, and the Americans diversified into many other aspects, from the impact of great cities, to the use of fire and the chilling story of world resource use in the twentieth century.

Donald Worster, the doyen of American environmental historians, in 1988 identified 'three sets of questions' that the subject seeks to answer, 'each drawing on a range of outside disciplines'.[10] First was 'understanding nature itself, as organized and functioning in past times', both in its organic and inorganic aspects. This sounds like Mei Xueqin's 'natural history' except that Worster explicitly includes 'the human organism as it has been a link in nature's food chains, now functioning as a womb, now belly, now eater, now eaten, now a host for microorganisms, now a kind of parasite'. Secondly comes 'the socioeconomic realm as it interacts with the environment'. He stressed more the impact of natural systems on human society than the impact of human society on natural systems, but it is the latter that strikes me as a peculiarly appropriate focus for our subject. Thirdly comes 'that more intangible and uniquely human type of encounter – the purely mental or intellectual, in which perceptions, ethics, laws, myths and other structures of meaning become part of an individual's or a group's dialogue with nature'. Geographers, using the preposterous language of post-modernism, are fond of saying that we 'construct' nature, though in reality they cannot even construct a wren. But we do vary in our attitudes towards nature and the meanings we attribute to the non-human world. Given our primacy as manipulators, this is an essential facet of environmental history. Worster rightly concludes that 'the new environmental history . . . brings together a

---

[9] For early Australian environmental history, see for example K. Hancock, *Discovering Monaro: a Study of Man's Impact on his Environment* (Cambridge, 1972); D. W. Meinig, *On the Margins of the Good Earth: the South Australian Wheat Frontier 1869–1884* (Chicago, 1962); M. Williams, *The Making of the South Australian Landscape* (London, 1974).

[10] D. Worster (ed.), *The Ends of the Earth: Perspectives on Modern Environmental History* (Cambridge, 1988), pp. 289–307.

wide array of subjects, familiar and unfamiliar, rather than setting up some new esoteric speciality'. Chapter 1 in this book carries this discussion further in asking how the British came to understand that the environment had a history and what is distinctive about the way we have examined it.

Many people, both students within the subject and policy makers and scientists outside it, will enquire what the relevance is of environmental history to modern environmental problems. Bad history is certainly not relevant and, just as social history was not served by idealogues cherry-picking facts and examples from the past to support a particular political point of view, so environmental history will not be served by a narrative that attributes all the misfortunes of the past, from the fall of Rome to the Black Death and Hurricane Katrina, to man's misuse of his environment. But environmental historians have perspectives of their own, and very often they do share the concerns of modern environmentalism, as I certainly do myself. Some of the essays in this volume (for example Chapters 12 and 13) were written against this background of concern, but I hope not in a manner that distorts their attempted objectivity.

Good environmental history is certainly relevant, simply because environmental change by definition is something that happens historically, over time, and to ignore a time dimension is to deprive its study of any context. In the great collaborative effort to understand the history of the seas under the direction of Poul Holm in Denmark, scientists and historians have come together to illuminate the crisis that has struck the world's largest ecosystem, more brilliantly than either could do alone.[11] If I was asked to name one single book that had most relevance to policy makers today, it would be J. R. McNeill's magisterial overview of the environmental history of the twentieth-century world, *Something New Under the Sun*, with its argument that the acceleration of resource use eclipses even the world wars and the rise and fall of Communism as the most important motif of the twentieth century.[12] If every student read it in his first year, our leaders might get an empowering perspective.

McNeill cites G. K. Chesterton at the head of his first chapter: 'the disadvantage of men not knowing the past is that they do not know the present'. That could serve as ample justification for all our labours.

---

[11] P. Holm, T. D. Smith and D. J. Starkey (eds), *The Exploited Seas: New Directions for Marine Environmental History* (St John's, Newfoundland, 2001); P. Holm, 'The last fish – a historical perspective on the exploitation of the North Sea', in T. Mizoguchi (ed.), *The Environmental Histories of Europe and Japan* (Nagoya, 2008), pp. 139–48.

[12] J. R. McNeill, *Something New Under the Sun: an Environmental History of the Twentieth-century World* (New York, 2000).

Fig. 1 John Alexander Harvie Brown (1844–1916), one of the most distinguished amateur naturalists in late Victorian Britain, author of pioneering histories of the capercaillie, red squirrel and great spotted woodpecker. The naturalist historian is a distinctive and enduring figure in British environmental history.

CHAPTER 1

The Environmental Historiography of Britain*

When did the British discover that the environment had a history? It was not in 1555, when the term 'natural history' entered the English language, or in 1686 when John Ray published the first volume of his *Historia Generalis Plantarum*, or even in 1771, when Gilbert White sketched to Thomas Pennant his project for 'a natural history of my native parish', to become the *Natural History of Selbourne*. 'History' as used in these titles meant 'description' in the Aristotelian sense of scholarly knowledge. Natural history was related closely to the respectful observation of nature expressed in the Book of Job and in the Psalms, recovered for Christianity by St Thomas Aquinas and St Francis, and expressed in English sculpture of the fourteenth century in the accurately observed foliage and animal carvings of Southwell Minster and the lively fishes swimming round the feet of the majestic Runcorn St Christopher, or in the birds depicted and named so lovingly in the Sherborne missal – nature as delight and as showing forth the glory of God.[1]

But natural history was essentially a Renaissance and a Baconian undertaking, a systematic account of a set of natural phenomena. It was divorced from the medieval fables of the bestiary, whereby creatures were set on earth to symbolise moral qualities, though this was to have a life of its own that flourished until Victorian times. It was not divorced, though, either from the modern project of discovering economic uses for nature, or from the ancient one of showing forth the glory of the Creator, later rechristened natural religion.[2] But natural history had no timeline, so it was not history in our sense.

---

* The text is slightly updated from a paper given in Coburg in September 2005 to a conference hosted by the Prinz-Albert-Gesellschaft on the theme 'Environment and History in Britain and Germany'. Thanks to K. G. Saur Verlag GmbH, München for permission to reproduce the text from *Prinz-Albert-Studien*, 24 (2006), 'Unwelt und Geschichte in Deutschland und Grossbritannien: Environment and History in Britain and Germany', pp. 37–46.

[1] See Nikolaus Pevsner, *The Leaves of Southwell* (London, 1945); Janet Blackhouse, *Medieval Birds of the Sherborne Missal* (London, 2001).

[2] Charles E. Raven, *English Naturalists from Neckam to Ray* (Cambridge, 1947); Charles E. Raven, *John Ray Naturalist. His Life and Works* (Cambridge, 1950); Keith Thomas, *Man and the Natural World: Changing Attitudes in England* (Harmondsworth, 1983).

The British discovered environmental history during the Enlightenment. As good a date as any is 1788, when the Scottish geologist James Hutton, friend of Hume and Smith, published his *Theory of the Earth* and concluded a scientific lecture given before the Royal Society of Edinburgh with the scalp-tingling words, 'we find no vestige of a beginning – no prospect of an end'. The notion that the world was only 6,000 years old, as reckoned from the Biblical generations, finally went out of the window then, though it had already been undermined – for instance by Newton's calculation that the molten sphere of iron that was the earth would take 50,000 years to cool. Even Hutton could not guess that the universe was about 13.7 billion years old, not 6,000 – that immensity would take until the twenty-first century to discover.[3]

Victorian geologists, who wanted both to remain Christian and to respect the science, of course knew through the study of strata and fossils that the environment was immensely ancient, and treated Genesis as a parable. One of the most influential was the very gifted Scottish amateur Hugh Miller, who argued that the seven biblical days of creation were in fact 'prophetic days' of long duration for which the periods of geology provided evidence: six earlier days of primitive sea-life, of primitive forests, of birds and reptiles, of mammals and of man. The present day is the seventh or Sabbath day, when man must be redeemed, and with the rise of regenerate man, God's intention for creation will be complete.[4] Miller was responding to Robert Chambers' early and faith-eroding views of evolution expressed in *Vestiges of the Natural History of Creation* (1844), but with Wallace and Darwin, evolution became continuous and needed no invisible hand at all.[5] By the time of the publication of the *Origin of Species* in 1858 the battle lines were drawn between the creationists and the evolutionists. There is a sense in which environmental history became the central intellectual problem of the Victorian age.[6]

---

[3] D. R. Dean, *James Hutton and the History of Geology* (Ithaca, 1992); Donald B. McIntyre and Alan McKirdy, *James Hutton, the Founder of Modern Geology*, National Museums of Scotland (Edinburgh, 2001).

[4] M. Shortland, *Hugh Miller and the Controversies of Victorian Science* (Oxford, 1996); Charles Waterston, 'An awakening interest in geology', in Lester Borley (ed.), *Hugh Miller in Context* (Cromarty, 2002), pp. 85–90. See also Lester Borley (ed.), *Celebrating the Life and Times of Hugh Miller* (Cromarty, 2003).

[5] James A. Secord, *Victorian Sensation, the Extraordinary Publication, Reception and Secret Authorship of Vestiges of the Natural History of Creation* (Chicago, 2000); Peter Raby, *Alfred Russel Wallace, a Life* (London, 2001); Peter J. Bowler, *Charles Darwin: the Man and His Influence* (Cambridge, 1990); Janet Browne, *Charles Darwin*, vol. 1: *Voyaging*, vol. 2: *The Power of Place* (London and New York, 1995 and 2002).

[6] Peter J. Bowler, 'Darwinism and Victorian values, threat or opportunity', in T. C. Smout (ed.), Victorian Values, a Joint Symposium of the Royal Society of Edinburgh and the British Academy (December, 1990), *Proceedings of the British Academy*, 78 (Oxford, 1992) pp. 129–47.

It was not, however, an environmental history that was happening around Victoria and Albert as they sat admiring the harvest fields on the Isle of Wight, or walked together in the ancient pine forest around Balmoral. It was the environmental history of the cosmos and the coal measures, of the fossil tree and the ammonite, the environmental history of physical geographers, which deals, or can deal, in thousands of millennia, and has no need for people to be on earth to have a history. 'Environmental history' is used in that sense by British physical geographers to this day, and remains a productive and brilliant tradition. For example, in 1996 Edinburgh University hosted a symposium on the Environmental History of the Cairngorms in which documentary history played no part at all, but which showed a dazzling array of scientific methodologies through which to explore the past. One paper used the analysis of the fossilised heads of non-biting midges deposited in high mountain lochs to indicate 3,000 years of climate change. Others used carbon and lead or zinc particles embedded in trees or deposited in water to study pollution over hundreds of years, and to show how changing levels of acidity had altered the diatom flora of fresh water, suddenly and abruptly in the twentieth century.[7]

I dwell on this example (which is not a particularly unusual one of the way geographers' work) to emphasise the way in which scientists over the last century have added an extraordinary range of tools to understand the environmental past, many of which ordinary historians are only dimly aware; in certain cases the scientists concerned have also combed documentary sources to add another, and what is for us more recognisable and recognised, perspective. Thus Peter Brimblecombe in his remarkable study of air pollution in London writes both as a scientist and a historian,[8] and the work of the climate history school at Norwich pioneered by H. H. Lamb involves ice-cores, historical records of temperature and rainfall since the seventeenth century and the qualitative evidence of diaries and chronicles.[9] Mike Baillie in Belfast argues for the value of dendrochronology, the study

---

[7] Botanical Society of Scotland, Symposium on Environmental History of the Cairngorms, *Botanical Journal of Scotland*, 48 (1) (1996). See in particular S. J. Brooks, 'Three thousand years of environmental history in a Cairngorms lochan revealed by analysis of non-biting midges', pp. 89–98; N. J. Loader and V. R. Switsur, 'Reconstructing past environmental change using stable isotopes in tree-rings', pp. 65–78; R. W. Battarbee, V. J. Jones *et al.*, 'Palaeolimnological evidence for the atmospheric contamination and acidification of high Cairngorm lochs, with special reference to Lochnagar', pp. 79–88.

[8] Peter Brimblecombe, *The Big Smoke. A History of Air Pollution in London since Medieval Times* (London, 1987).

[9] H. H. Lamb, *Climate: Present, Past and Future* (London, 1972); *Climate, History and the Modern World* (London, 1982).

of tree-rings, not only as an archaeological dating technique but also as a source for the history of climatic crises over at least two millennia.[10] Jean Grove's classic study of the Little Ice Age is a synthesis of much European as well as British climate history research in this tradition.[11]

Of all the methodological contributions that environmental scientists have made, probably none has greater current significance for the study of environmental history than radiocarbon dating, used alongside palynology, the study of fossil pollen, two fields in which the main pioneer in Britain was Sir Harry Godwin of Cambridge University, who applied the techniques to develop the theory of ecological succession. His work is commemorated and furthered in the Godwin laboratory there. While early radiocarbon dating was relatively crude, recent advances in that and allied dating techniques, using lead isotopes, have led to considerable refinement.

Palynology can be applied to previous interglacials where the muds and peats are old enough, but its greatest application has been to study the early and mid Holocene (the period since the last Ice Age), and through it we have gathered a fairly clear idea of the succession of forest communities in Britain and Europe, that formed the environmental setting of the Mesolithic, the Neolithic, the Bronze Age and the Iron Age. It has helped archaeologists and early historians to revise completely their view of how wooded Scotland was, for example, at the time of the Roman invasion. In this context I am most cognisant of the work of Richard Tipping of Stirling, whose work on Scottish woodland history, and that of his students, has been of outstanding value to the historical and archaeological community.[12]

A very recent example of the use of palynology, and of that of the comparable study of early beetle fauna from insect remains, has been to test Franciscus Vera's hypothesis that the original-natural wildwood of the postglacial Mesolithic millennia had not been closed forest but open country in which the woods were groves surrounded by thorny edges, with grazing by large herbivores as the force behind cyclical vegetation dynamics and shifting mosaics of forest. The weight of evidence from the scientists assembled by Keith Kirby for English Nature has been to suggest that Vera's hypothesis, as stated, is overdrawn, and that Britain had a mixed landscape with

---

[10] M. G. L. Baillie, *A Slice Through Time: Dendochronology and Precision Dating* (London, 1995).

[11] Jean Grove, *The Little Ice Age* (London, 1988).

[12] Richard Tipping, 'The form and fate of Scottish woodlands', *Proceedings of the Society of Antiquaries of Scotland*, 124 (1994), pp. 1–54; 'Living in the past, woods and people in prehistory to 1000 BC', in T. C. Smout (ed.), *People and Woods in Scotland* (Edinburgh, 2003), pp. 14–39. See also C. Dickson and J. Dickson, *Plants and People in Ancient Scotland* (Stroud, 2000).

some areas of vegetation cycles and others with more permanent closed forest, with herbivores not always the driving force.[13]

Relatively little palynology has seriously overstepped the boundary of the last 1,000 years except in one or two recent Ph.D. theses, but it is fascinating for the documentary historian to see what these techniques might yield for the relatively recent past.[14]

To palynology and all those other techniques I have mentioned should be added the work of soil scientists, who seem to me to have it within their power one day, through the analysis of buried soils, to be able to throw much light not only on early manuring practices but also, for example, on questions of nutrient exhaustion in medieval agriculture.[15] Similarly, genetic studies and the science of DNA have potential to illuminate historical patterns of human and animal migration in ways at which we can barely guess.

I have emphasised the contributions of applied physical science and physical geography because I believe that all modern environmental historians can still learn so much from it. But modern environmental history has a variety of other roots, one of which is the interest that zoologists themselves felt in the way that their subjects of study had interacted with man in recent centuries. Victorian natural historians were increasingly interested in species history, especially where man had brought about or threatened extinction. The extinction of the great auk was traumatic to Victorian sensibilities and led the Cambridge zoologist Alfred Newton down the road towards legislative protection for seabirds, and it also led in 1885 to a remarkable species history of the great auk by Symington Grieve.[16] J. H. Gurney of Norwich wrote on the gannet, 'a bird with a history', and the Scottish naturalist J. A. Harvie Brown wrote a series of species histories of taxa either at risk from local extinction or re-introduced after extinction, the capercaillie, the great spotted woodpecker and the red squirrel, the last of which led to a book that aroused the special attention of the Royal

---

[13] F. W. M. Vera, *Grazing Ecology and Forest History* (Wallingford, 2000); K. H. Hodder, J. M. Bullock, P. C. Buckland and K. J. Kirby, *Large Herbivores in the Wildwood and Modern Naturalistic Grazing Systems*, English Nature Report 648 (2005).

[14] Philip Sansum, 'Historical resource use and ecological change in semi-natural woodland: western oakwoods in Argyll, Scotland', University of Stirling Ph.D. thesis, 2004.

[15] E. I. Newman and P. D. A. Harvey, 'Did soil fertility decline in Medieval English farms? Evidence from Cuxham, Oxfordshire, 1320–1340', *Agricultural History Review* 45 (1997), pp. 119–136; D. A. Davidson and I. A. Simpson, 'Soils and landscape history – case studies from the Northern Isles of Scotland', in S. Foster and T. C. Smout (eds), *The History of Soils and Field Systems* (Aberdeen, 1994), pp. 66–74.

[16] Symington Grieve, *The Great Auk or Garefowl (Alca impennis Linn.): its History, Archaeology and Remains* (Edinburgh, 1885). More recent studies include Errol Fuller, *The Great Auk* (Southborough, Kent, 1999); W. R. P. Bourne, 'The story of the great auk *Pinguinus impennis*', *Archives of Natural History*, 20 (2) (1993), pp. 257–78.

Family.[17] When natural history turned towards ecology, an interest grew in the interaction between biological groups or communities and man. It was evident early in the twentieth century by interest in the German concept of 'urvald' or wildwood, and how far it might apply in Britain.[18] After the First World War, James Ritchie's *Influence of Man on Animal Life in Scotland* (Cambridge, 1920) was a milestone, ambitious and comprehensive, very modern in conception and execution but still under-estimated.

After the Second World War, the tradition continued, especially with the Collins New Naturalist series. This encompassed a series of volumes all by scientists, who, like Ritchie, were caught up in the emerging nature conservation movement and wished to write about the interactions between man and nature, inevitably with an historical perspective: R. S. R. Fitter, *London's Natural History* (London, 1945), E. M. Nicholson, *Birds and Men: the Birdlife of British Towns, Villages, Gardens and Farmland* (London, 1951) and H. L. Edlin, *Trees, Woods and Man* (London, 1956), are good examples of this type of study.[19] It is a tradition that continues to provide distinguished work, as in Derek Yalden, *The History of British Mammals* (London, 1999) and Michael Shrubb, *Birds, Scythes and Combines: a History of Birds and Agricultural Change* (Cambridge, 2003), which provides a detailed and perceptive study of the interaction of agricultural history and birdlife since the eighteenth century. In the same lineage is Roger Lovegrove, *Silent Fields* (Oxford, 2007), a history of the human treatment of birds in Britain, centred on exhaustive research of English church-wardens' accounts of 'vermin' bounties since the reign of the Tudors, and Bryony Coles, *Beavers in Britain's Past* (Oxford, 2006). Also of great interest is the floral history of Glasgow produced by Jim Dickson and his collaborators, showing how vibrant and dynamic the ecosystem of a great city can be, no whit less interesting for the role of so many despised 'alien species'.[20]

The outstanding example of the tradition of scientist working as historian, however, is Oliver Rackham, probably the only environmental histo-

---

[17] J. H. Gurney, *The Gannet: a Bird with a History* (London, 1913); J. A. Harvie Brown, *The Capercaillie in Scotland* (Edinburgh, 1879); *The History of the Squirrel in Great Britain* (Edinburgh, 1881); 'The great spotted woodpecker *(Picus major, L.)* in Scotland', *Annals of Scottish Natural History*, 1 (1892), pp. 4–17.

[18] G. P. Gordon, 'Primitive woodland and plantation types in Scotland', *Transactions of the Royal Scottish Arboricultural Society*, 24 (1911), pp. 153–77.

[19] See P. Marren, *The New Naturalists* (London, 2005); though by a human geographer rather than a scientist, another book in the same series is also environmental history: L. Dudley Stamp, *Man and the Land* (London, 1964). It contains among other things a chapter on alien species.

[20] J. H. Dickson, P. Macpherson and K. Watson (with contributions from others), *The Changing Flora of Glasgow. Urban and Rural Plants through the Centuries* (Edinburgh, 2000).

rian whose name is widely recognisable to the contemporary British public. Rackham is a Cambridge botanist by training who has acquired the skills of a medieval and early modern historian, and is a Fellow of the British Academy rather than of the Royal Society on account of his contribution to the humanities. His works on woodland history, of the Mediterranean as well as of England, have had international impact, and by defining and emphasising the concept of ancient woodland as a living category in the English countryside, he has had a profound practical impact on modern British conservation. Scarcely any forester in the British Isles can now think of his job without being aware of woods as historical artefacts and of the work of Oliver Rackham in sensitising the public to their history.[21] He began his notable studies with detailed work on a local wood in Cambridgeshire, *Hayley Wood: its History and Ecology* (Cambridge, 1975), but his most influential works on British woodland history have been *Trees and Woodland in the British Landscape: the Complete History of Britain's Trees, Woods and Hedgerows* (London, 1976), *Ancient Woodland, its History, Vegetation and Uses in England* (London, 1980: new edn Colvend, Kirkudbrightshire, 2003) and, more recently, his New Naturalist volume, *Woodlands* (London, 2006). It was, however, his *History of the Countryside* (London, 1986), a popular and general but never condescending work, that first excited me about the intellectual possibilities of environmental history.

It is interesting that Rackham has been a consistent critic of the sort of environmental history that lays more emphasis on the history of people using and thinking about nature than about the history of natural things used by people: as he accurately put it concerning a volume that I edited, a book entitled *People and Woods* is more likely to be mainly about people than mainly about woods. It is fair comment, but since the 1970s various sorts of British historians, and no longer just scientists, have interested themselves in environmental history, and they do also tend to be seriously interested in people. The rise of a new political and intellectual 'environmentalism' inspired some to explore the origin of changes in attitudes towards nature and the rise of conservation politics. Keith Thomas, polymath among English early modern historians, produced his outstanding and influential volume, *Man and the Natural World* (Harmondsworth, 1983), tracing how English ideas about plants and animals developed between 1500 and 1800. To a degree, it had been paralleled and anticipated in literary studies a decade earlier by Raymond Williams's equally brilliant *The Country and the City* (London,

---

[21] Judith Tsouvalis, *A Critical Geography of Britain's State Forests* (Oxford, 2000); J. Tsouvalis-Gerber, 'Making the invisible visible: ancient woodlands, British Forest Policy and the Social Construction of Reality', in Charles Watkins (ed.), *European Woods and Forests, Studies in Cultural History* (Wallingford, 1998).

1973). John Sheail, trained as a Tudor historian but working for the Nature Conservancy Council, produced a range of seminal volumes about the history of ecology conservation and environmental planning in Great Britain.[22] D. E. Allen wrote a memorable history of British naturalists in 1976, emphasising in particular the distinctive national amateur tradition.[23]

Others, notably economic historians and human geographers, began to examine resource history in the light of environmental concerns and interests. In 1994, Brian Clapp wrote a sketchy but highly pioneering overall view of the environmental history of Britain since the Industrial Revolution, albeit from a rather narrow economic historian's point of view.[24] More ambitious was the work of Ian Simmons, who as a geographer in 2001 capped many years of scholarship exploring environmental history (from the archaeological perspective among others) with a magisterial overview of British environmental history as a matter of resource use.[25] Angus Winchester explored the environmental history of the northern English and southern Scottish uplands very much with the eye of an economic and agrarian historian interested in allocation and community division of resources.[26]

At Oxford, the annual Linacre lectures on the environment since 1997 have resulted in seven volumes which surprisingly seldom deal with environmental history and still less often with the British experience of it.[27] On the other hand, the Institute for Environmental History at St Andrews produced, between 1993 and 2002, a series of books based on conference proceedings, ranging from soil history and species history to landscape history, mainly, though not exclusively, with a Scottish emphasis.[28] In the

---

[22] John Sheail, *Nature in Trust, the History of Nature Conservation in Britain* (London and Glasgow, 1976); *Rural Conservation in Interwar Britain* (Oxford, 1981); *Pesticides and Nature Conservation: the British Experience 1950–1975* (Oxford, 1985); *Seventy-five Years in Ecology, the British Ecological Society* (Oxford, 1987); *Nature Conservation in Britain – The Formative Years* (London, 1998); *An Environmental History of Twentieth-Century Britain* (Basingstoke, 2002).

[23] David Elliston Allen, *The Naturalist in Britain, a Social History* (London, 1976).

[24] Brian W. Clapp, *An Environmental History of Britain since the Industrial Revolution* (London, 1994).

[25] I. G. Simmons, *An Environmental History of Great Britain from 10,000 Years Ago to the Present* (Edinburgh, 2001). See also Ian Simmons and Michael Tooley (eds), *The Environment in British Prehistory* (Cornell, 1981).

[26] Angus J. L. Winchester, *The Harvest of the Hills, Rural Life in Northern England and the Scottish Borders, 1400–1700* (Edinburgh, 2000). See also I. G. Simmons, *The Moorlands of England and Wales, an Environmental History 8000 BC–AD 2000* (Edinburgh, 2003).

[27] The most relevant are Kate Flint and Howard Morphy (eds), *Culture, Landscape and the Environment* (Oxford, 2000), and Paul Slack (ed.), *Environments and Historical Change* (Oxford, 1999).

[28] T. C. Smout (ed.), *Scotland Since Prehistory: Natural Change and Human Impact* (Aberdeen, 1993); S. Foster and T. C. Smout (eds), *The History of Soils and Field Systems* (Aberdeen, 1994); Graeme Whittington (ed.), *Fragile Environments, the Use and Management of*

Ford lectures in Oxford in 1999 I tried to suggest a tension between use of nature and a delight in nature, in the specific context of the environmental history of Scotland and Northern England since 1600.[29] The same theme was taken up by Robert Lambert in a study of conflict over land use in the Cairngorms from 1880–1980.[30] Others, notably Charles Watkins, Richard Muir, William Linnard, Ian Rotherham and members of the Scottish Woodland History Discussion Group examining the history of native woods, built on the work of Rackham to examine new areas of woodland history both in the geographical and intellectual sense.[31] Stephen Moseley followed Brimblecombe's lead with a study of smoke pollution in Victorian and Edwardian Manchester, but with more focus on social and political urban history.[32]

So far, perhaps surprisingly to the reader, I have dealt without reference to the influence of the great efflorescence of environmental history that occurred in America from the 1970s onwards, which has produced such notable authors as Donald Worster, William Cronon, J. R. McNeill, Roderick Nash, Caroline Merchant, Stephen Pyne, Alfred Crosby and Jahred Diamond.[33] The list is much longer than I can indicate here, and in

---

Tentsmuir NNR, Fife (Edinburgh, 1997); T. C. Smout (ed.), *Scottish Woodland History* (Edinburgh, 1998); Robert A. Lambert (ed.), *Species History in Scotland, Introductions and Extinctions since the Ice Age* (Edinburgh, 1998); T. C. Smout and R. A. Lambert (eds), *Rothiemurchus: Nature and People on a Highland Estate, 1500–2000* (Dalkeith, 1999); T. C. Smout (ed.), *Understanding the Historical Landscape in its Environmental Setting* (Dalkeith, 2002).

[29] T. C. Smout, *Nature Contested, Environmental History in Scotland and Northern England since 1600* (Edinburgh, 2000).

[30] Robert A. Lambert, *Contested Mountains, Nature, Development and the Environment in the Cairngorms Region of Scotland, 1880–1980* (Cambridge, 2001).

[31] Keith J. Kirby and Charles Watkins (eds), *The Ecological History of European Forests* (Wallingford, 1998); Charles Watkins (ed.), *European Woods and Forests, Studies in Cultural History* (Wallingford, 1998); Richard Muir, *Ancient Trees, Living Landscapes* (Stroud, 2005); William Linnard, *Welsh Woods and Forests, a History* (Llandysul, 2000); P. Beswick and I. Rotherham (eds), *Ancient Woods, their Archaeology and Ecology* (Sheffield, 1993); T. C. Smout (ed.), *People and Woods in Scotland, a History* (Edinburgh, 2003); T. C. Smout, Alan R. MacDonald and Fiona Watson, *A History of the Native Woodlands of Scotland, 1500–1920* (Edinburgh, 2005). The Scottish Woodland History Discussion Group Notes have been published annually, based on conference papers, since 1996 (ISSN 1470-0271). In Scotland there is an independent older tradition of woodland history written by foresters going back to H. M. Steven and A. Carlisle, *The Native Pinewoods of Scotland* (Edinburgh, 1959) and Mark Louden Anderson, *A History of Scottish Forestry*, 2 vols (London, 1967).

[32] Stephen Moseley, *The Chimney of the World: a History of Smoke Pollution in Victorian and Edwardian Manchester* (Cambridge, 2001).

[33] The following is a mere sample of their work: Donald Worster, *Nature's Economy, a History of Ecological Ideas* (Cambridge, 1977); William Cronon, *Nature's Metropolis: Chicago and the Great West* (New York, 1991); J. R. McNeill, *Something New Under*

many respects the American school dominates the subject, so that today what is usually meant by 'environmental history' internationally is the humanities-based and often ideologically committed subject associated with the Americans. Of course it has had enormous influence in Britain as elsewhere, but most notably it has been among historians working in British universities but studying other continents: Richard Grove, for instance, on tropical islands and on India, William Beinart and John MacKenzie on Africa, Mark Elvin on China, Peter Coates on America, to name only some of the most distinguished.[34] This is mainly because, as defined by the American school, environmental history has essentially been an extension of the history of imperial expansion, of colonial impacts with European crops, animals and machinery on fragile environments, whether these be the Arctic, the mid-western plains, Caribbean islands, the tropics or the South African veldt. The questions they were asking seemed much less appropriate to a country (indeed, to a continent) where invasion and overthrow are not usually seen in the same imperial expansionist terms, and where the environment both appears superficially to be much more robust (at least in terms of soil and water) and where, perhaps more pertinently, crops and animals evolved over centuries to grow, more or less sustainably, within it.

The history of the environment in Britain seemed less to be environmental history in the American sense of a history of misunderstanding and violent misuse, as a history of relatively benign and gradual landscape change and of an agriculture that, by certain definitions at least, was environmentally sustainable until after the Second World War. Landscape history and agrarian history are subjects in which British geographers and historians have

---

the Sun: an Environmental History of the Twentieth-Century World (New York, 2000); Roderick Nash, Wilderness and the American Mind, 3rd edn (New Haven, 1982); Carolyn Merchant, Ecological Revolutions, Nature, Gender and Science in New England (Chapel Hill, 1989); Stephen J. Pyne, Fire in America, a Cultural History of Wildland and Rural Fire (Princeton, 1982); Alfred W. Crosby, Ecological Imperialism, the Biological Expansion of Europe, 900–1900 (Cambridge, 1986); Jared Diamond, Collapse, How Societies Choose to Fail or Survive (London, 2005).

34 Richard H. Grove, Green Imperialism: Colonial Expansion, Tropical Island Edens and the Origins of Environmentalism, 1600–1960 (Cambridge, 1995); William Beinart, The Rise of Conservation in South Africa: Settlers, Livestock and the Environment, 1770–1850 (Oxford, 2003); John M. MacKenzie, The Empire of Nature: Hunting, Conservation and British Imperialism (Manchester, 1988); Mark Elvin, The Retreat of the Elephants, an Environmental History of China (London, 2004); Peter A. Coates, The Trans-Alaska Pipeline Controversy: Technology, Conservation and the Frontier (Fairbanks, 1991); William J. Beinart and Peter A. Coates, Environment and History: the Taming of Nature in the USA and South Africa (London, 1995); Peter A. Coates, American Perceptions of Immigrant and Invasive Species (Berkeley, 2006). This is also only a sample of the works of these authors.

been distinguished since the days of H. C. Darby and W. G. Hoskins,[35] a tradition which is continued today by cultural geographers like David Lowenthal and historical geographers like Robert Dodgshon.[36] As might be expected, geographers as a profession have had a particular sensitivity to environmental history, exemplified by two volumes published by the Institute of British Geographers in 1976 and 1995.[37] Journals on landscape history[38] also often contain more British environmental history than the two periodicals explicitly devoted to environmental history, the American-based *Environmental History* and the British-based *Environment and History*. Neither could be said to discriminate against British material, yet they fail to attract it as often as could be expected. The other route is the kind of economic history exemplified in the multi-volume *Agrarian History of England and Wales from Pre-history to 1939*, under the general editorship of Joan Thirsk, providing an achievement of truly monumental scholarship.[39] Joan Thirsk's own work, for example her history of alternative agriculture in England since the Black Death, could be considered an example of environmental history as much as economic history, so much do the two interweave and interlock.[40]

However, the sole example I can think of where a historian explicitly asks how fragile was the British environment in the face of industrial and agrarian change is James Winter's fine study of the impact on the Victorian countryside of the huge contemporary changes in economy and society.[41] Significantly, Winter's home base is the University of British Columbia in Vancouver. Yet Winter's book seems to be virtually unknown to mainstream scholars of nineteenth-century Britain. This points to a further weakness that

[35] H. C. Darby, *The Domesday Geography of England*, 7 vols (Cambridge, 1952–77); W. G. Hoskins, *The Making of the English Landscape* (London, 1955).

[36] David Lowenthal, *The Past is a Foreign Country* (Cambridge, 1985); E. C. Penning-Rowsell and David Lowenthal (eds), *Landscape Meaning and Values* (London, 1986); Robert A. Dodgshon, *Land and Society in Early Scotland* (Oxford, 1981), *Society in Time and Space: a Geographical Perspective* (Cambridge, 1998), 'Budgeting for survival, nutrient flow and traditional Highland farming', in S. Foster and T. C. Smout (eds), *The History of Soils and Field Systems* (Aberdeen, 1995), pp. 83–93.

[37] L. F. Curtis and I. G. Simmons (eds), *Man's Impact on Past Environments*. Transactions of the Institute of British Geographers, n.s. 1 (1976), no. 3; Robin A. Butlin and Neil Roberts (eds), *Ecological Relations in Historical Times*, Special Publications 32 (Oxford, 1995).

[38] The journals are *Landscape History* and *Landscapes*. The *Agricultural History Review* also often contains articles of relevance to British environmental history.

[39] Joan Thirsk (gen. ed.), *Agrarian History of England and Wales*, 8 vols (Cambridge, 1967–2000).

[40] Joan Thirsk, *Alternative Agriculture. A History from the Black Death to the Present Day* (Oxford, 1997).

[41] James Winter, *Secure from Rash Assault. Sustaining the Victorian Environment* (Berkeley, 1999).

Paul Warde and Sverker Sörlin have recently urged us to consider, namely that most environmental historians seem to live in a ghetto where they do not interact much with other historians, and they are largely ignored by them in turn.[42]

Warde's own recent work on the history of energy consumption since 1560, though, is a brilliant example of a book that can only be ignored by mainstream historians at their peril, because its conclusions relating to the date at which England became locked into coal burning, and the careful quantitative study of the rise of coal and then oil to total domination, are so important.[43] Indeed, the problem of the history of energy use, first conceptualised by the economic historian Tony Wrigley as the transition from an organic to an inorganic economy, and then the subject of an inspiring pan-European seminar at Prato in 2002, has created a new interface between environmental and economic history where probably the most exciting advances in either discipline are taking place.[44] International cooperation, combined with the rigorous quantification that Warde exemplifies in Britain, Malanima in Italy and Kander in Sweden, will be immensely fruitful.[45]

Another interesting question is the relationship between urban history and environmental history in Britain. When Bill Luckin writes about water pollution in the nineteenth-century Thames or Stephen Moseley about air pollution in nineteenth-century Manchester, is it urban or environmental history?[46] At my own university of St Andrews John Clark's project on the history of urban waste is labelled and funded as environmental history but it could just as easily have been funded as urban history or social history. Perhaps this ambiguity is a strength, after all, and could help to overcome the ghetto mentality by a kind of permeability with other forms of history.

What of the future? I can sum up my impressions in three sentences. Firstly, we should celebrate that such a lot is going on especially among

---

[42] S. Sörlin and P. Warde, 'The problem of the problem of environmental history – a re-reading of the field and its purpose', *Environmental History*, 12 (2007), pp. 107–30.

[43] P. Warde, *Energy Consumption in England and Wales, 1500–2000* (Consiglio Nazionale della Ricerche, Naples, 2008), ISBN 978-88-8080-082-8.

[44] E. A. Wrigley, *Continuity, Chance and Change: the Character of the Industrial Revolution in England* (Cambridge, 1988); E. A. Wrigley, 'The transition to an advanced organic economy; half a millennium of English agriculture', *Economic History Review*, 59, pp. 435–80; S. Cavaciocchi (ed.), *Economia e Energia, secc. XIII–XVIII* (Florence, 2003).

[45] P. Malanima, *Energy Consumption in Italy in the 19th and 20th Centuries: a Statistical Outline* (Consiglio Nazionale della Ricerche, Naples, 2008); A. Kander, *Economic Growth, Energy Consumption and CO2 Emissions in Sweden 1800–2000* (Lund, 2002).

[46] B. Luckin, *Pollution and Control: a Social History of the Thames in the Nineteenth Century* (Bristol, 1986); Moseley, *Chimney of the World*.

young scholars, Ph.D. students and lecturers, in however disparate a fashion, because out of all this intellectual ferment new ideas and concepts will spring more readily than if we were all old, over-focused and self-constrained. Secondly, we should not worry that the environmental history of Britain (or of Europe) is not necessarily about the same things as that of America, Australia or India, but, as Winter and Warde have shown, and others are taking on board, we could certainly learn to ask more questions about environmental sustainability.[47] Lastly, we should build much more on interdisciplinary cooperation with our scientific colleagues, because it is here especially that British environmental history has actually been at its most pioneering and valuable, not just since the 1970s, but from Victorian times.

---

[47] See www.historyand.sustainability.org, a project of the Centre for History and Economics, King's College, Cambridge.

Fig. 2 Lochdhu Lodge, Caithness, viewed over the Flow Country in the early 1980s before this area was drained and planted with Sitka spruce. To one set of eyes, this was a wasteland waiting to be developed; to another, a landscape and a habitat of international importance. Photographer: Glyn Satterley. Reproduced by kind permission.

CHAPTER 2

The Highlands and the Roots of Green Consciousness, 1750–1990*

I am returned from Scotland, charmed with my expedition: it is of the Highlands I speak: the Lowlands are worth seeing once, but the mountains are ecstatic and ought to be visited in pilgrimage once a year. None but those monstrous creatures of God know how to join so much beauty with so much horror. A fig for your poets, painters, gardeners and clergymen, that have not been among them, their imagination can be made up of nothing but bowling greens, flowering shrubs, horse-ponds, Fleet ditches, shell grottoes and Chinese rails. Then I had so beautiful an autumn. Italy could hardly produce a nobler scene, and this is so sweetly contrasted with that perfection of nastiness and total want of accommodation that only Scotland can supply.[1]

The words are Thomas Gray's, the year 1765, five years after the publication of Macpherson's *Ossian* had begun to form a taste among the southern educated for the gloomy splendours of the north. Although it was all so different from the gentle, pastoral, inhabited world, celebrated in his own polished verses, as far as I know he was the first Englishman to express admiration for Highland scenery. His visit fell between those of others more typical in their opinions: Thomas Burt, the Hanoverian officer, was one such. His own admitted notion of a 'poetical mountain' was Richmond Hill, and he found 'the monstrous excrescences' of the Highlands simply appalling – 'and the whole of a dismal gloomy brown drawing upon a dirty

---

* This essay originated as the Raleigh Lecture on British History, delivered at the University of Glasgow on 24 October 1990 and again at the British Academy in London on 20 November 1990. The text has been only slightly amended, and it should be seen in the context of British conservation politics at the end of the twentieth century. A short postscript has, however, been added to take account of criticism of the lecture. Thanks to the British Academy for permission to reproduce the text, amended from *The Proceedings of the British Academy*, 76 (1991), pp. 237–64.

[1] J. Mitford (ed.), *The Correspondence of Thomas Gray and William Mason* (London, 1853), p. 349.

purple; and most of all disagreeable when the heath is in bloom.'[2] Another was Samuel Johnson, also observing the chaos of the hills with an eye of a man more comfortable in a London coffee house: 'matter incapable of form or usefulness: dismissed by nature from her care ... quickened only with one sullen power of useless vegetation'.[3]

Gray we may then regard as an early swallow foretelling the summer of the aesthetic appreciation of the Highlands by outsiders, part of a European movement of exploring the sublime and the picturesque that began elsewhere.[4] It would not be true to say that, before the eighteenth century, mountains had been entirely without admirers.[5] Nor would it be true to say that a delight in nature was necessarily unknown before this period, or imported by outsiders. The Scottish Renaissance poets had a sharp ear for birdsong, for example, and the Gaelic poet Duncan Ban MacIntyre, himself a gamekeeper, an exact contemporary of Gray, published in 1768 what some have considered the finest nature poem in any British language. His praise of the hills struck another note from Burt and Johnson:

> In flawless green raiment
> as bright as the diamond
> your blooms in agreement
> like elegant music.[6]

It is from roots like these that an indigenous Scottish and Highland green consciousness could be traced. Even Burt admitted that the Highlanders themselves did not entirely share his prejudice: 'And, certainly, it is the deformity of the hills that makes the natives conceive of their naked straths and glens, as of the most beautiful objects in nature.'[7]

---

[2] *Burt's Letters from the North of Scotland, 1754* (Edinburgh, 1998), pp. 157–8.

[3] Samuel Johnson, *A Journey to the Western Islands of Scotland in 1773* (London, 1876), p. 32.

[4] S. H. Monk, *The Sublime: a Study of Critical Theories in XVIII century England* (New York, 1935); T. C. Smout, 'Tours in the Scottish Highlands from the eighteenth to the twentieth centuries', *Northern Scotland*, 5 (1983), pp. 99–121; P. Womack, *Improvement and Romance: Constructing the Myth of the Highlands* (London, 1989).

[5] Petrach, for instance, climbed the hills of Provence with delight, but rebuked himself by recalling St Augustine's warning that those who admired mountains, rivers and seas were admiring what was earthly and distracting themselves from the proper contemplation of God. See his letter of 26 April 1336 to Francesco Dionigi, reproduced in E. Cassirer, P. O. Kristeller and J. H. Randall (eds), *The Renaissance Philosophy of Man* (Chicago, 1948), pp. 36–46.

[6] There is excellent treatment of this problem in John Veitch, *The Feeling for Nature in Scottish Poetry* (Edinburgh, 1887). For the fifteenth and sixteenth century poets, see vol. 1, pp. 97–107. See also J. C. Shairp, *On the Poetic Interpretation of Nature* (Edinburgh, 1877); Duncan Ban Macintyre, *Ben Dorain*, translated from the Gaelic by I. C. Smith (Preston, 1969).

[7] *Burt's Letters*, p. 158.

Nevertheless, the eighteenth century did most emphatically see an alteration in consciousness among polite and educated southrons about the relationship between man and the natural world, as Keith Thomas in particular has eloquently described.[8] Up to this point, their attitude to the Highlands had been entirely utilitarian. Woods were cut and planted, ores were mined, land was farmed, seas were fished with no eye except for profit or subsistence. There then arose an appreciation of the Highlands as a place to visit for recreation – hence a tourist trade, and as a place to value for its own sake – hence restrictions on land use. By 1982, 30 per cent (and by 1990, 37 per cent) of the local government area Highland Region had been designated as being nationally important for nature and/or landscape conservation; it contained 40 per cent of all the land that was British National Nature Reserve (NNR), and 54 per cent of Scottish scenic areas.[9] It appeared that between a quarter and a third of the entire surface of the Highlands and Islands in the wider geographical sense was by then either designated or under consideration for designation for NNR, National Park Direction Area, or Site of Special Scientific Interest (SSSI); the Highlands were declared the only British priority area in the world conservation strategy.[10] So very tense had the situation become between traditional land users, developers of various kinds – including those in favour of tourist development – and conservationists, that in 1989 the Secretary of State for Scotland connived in the break-up of the British Nature Conservancy Council (NCC; which was responsible to the Secretary of State for the Environment in London), and proposed to assume direct command by creating a Scottish environmental agency responsible to himself in Edinburgh.

The purpose of this lecture is to examine the context of a movement, late, slow and incomplete though it may have been, to conserve the Highlands from depredation upon its scenery and ecology. The story, I suggest, touches upon more than may appear at first sight. It encapsulates a conflict that in rural areas over much of the western world is becoming critical at the end of the twentieth century, between growth – as envisaged in terms of income and employment generated by the traditional land uses of farming, fishing and forestry (the Highlands as private property) – and aesthetics – a way of looking at the Highlands as a place for recreation and preservation (the Highlands

---

[8] Keith Thomas, *Man and the Natural World: Changing Attitudes in England 1500–1800* (London, 1983).

[9] Highland Regional Council, *Conference on the Economic Future of the Highlands* (Inverness, 1982), p. 2; Countryside Commission for Scotland, *The Mountain Areas of Scotland: Conservation and Management* (1990) p. 15.

[10] J. M. Bryden in *Conference on the Economic Future*, p. 37. The figures have probably not changed much, but 'National Park Direction Area' is now an obsolete designation and 'National Scenic Area' needs to be included.

as heritage, a harmless-sounding word not far from the demon of public property). We are involved at present in a scene of considerable and bitter controversy, often of a complex kind, as anyone acquainted with the disputes of the 1980s will know: over Duich Moss, on Islay, over the Flow Country of Caithness (involving users of peat and potential forest land versus conservationists) or over the Northern Corries of Cairngorm (involving ski-developers versus conservationists). How did we get from there to here?

The Victorians adored the Highlands but did little or nothing to protect them. Ecologists and ecological historians have still only a shadowy picture of the impact in biological terms of the Highland clearances and the intrusion of outside market demands. In a series of writings between 1938 and 1968, Frank Fraser Darling taught what he believed to be the dire effect on the ancient forests, already damaged by Viking and medieval depredations, of the subsequent activities of English ironmasters in the seventeenth and eighteenth centuries and the sheep ranching that followed in the nineteenth century, resulting in the reduction of the Highlands to what he termed, in a vivid phrase, 'a wet desert'.[11] We should be cautious about accepting this dramatic picture in quite these terms. For one thing, most of the ancient forest that in late Mesolithic times covered perhaps 50–60 per cent of the land surface had already disappeared as a result of climatic change and peasant use of wood pasture and tree clearance over millenia.[12] Furthermore, what was left was more likely to have been preserved than damaged as the result of exploitation for charcoal and tanbark, as the ironmasters and the bark peelers had a reason for enclosing and looking after the woods to ensure sustainable resources; oakwoods (such as those of the Argyll sea lochs or Loch Lomondside) resumed their decline only in the nineteenth century when coal and chemicals replaced the forest product.[13]

---

[11] Fraser Darling refers to disafforestation in many places in his writings, usually with recognition that the process was long and drawn out, but often with the implication that most of the serious damage occurred after 1600. See, for instance, *Wild Country: a Highland Naturalist's Notes and Pictures* (Cambridge, 1938), pp. 69ff.; *The Story of Scotland* (London, 1942), p. 36; *Island Farm* (London, 1943), pp. 200–1; *West Highland Survey* (Oxford, 1955), pp. 3–5, 160–71; *Natural History in the Highlands and Islands* (London, 1947), reissued and revised as F. Fraser Darling and J. Morton Boyd, *Highlands and Islands* (London, 1964), pp. 55–62; *Pelican in the Wilderness* (London, 1956), pp. 180, 203; *Wilderness and Plenty, the Reith Lectures, 1969* (London, 1970), pp. 21–3. Much of his picture relies on the portrayal of disafforestation in James Ritchie, *The Influence of Man on Animal Life in Scotland* (Cambridge, 1920).

[12] T. C. Smout, A. R. MacDonald and F. Watson, *A History of the Native Woodlands of Scotland, 1500–1920* (Edinburgh, 2005), chs 2 and 3.

[13] Smout, MacDonald and Watson, *Native Woodlands*, ch. 7; J. M. Lindsay, 'The history of oak coppice in Scotland', *Scottish Forestry*, 29 (1975), pp. 87–95; R. M. Tittensor, 'History of Loch Lomond oakwoods', *Scottish Forestry*, 24 (1970), pp. 100–18. See also ch. 5 below.

The ancient pinewoods were already of limited extent in the seventeenth and eighteenth centuries, threatened by overgrazing by increasing numbers of black cattle (not sheep) and a rising density of human population – which reached its maximum in the Highlands and Islands in the census of 1841. They underwent dramatic decline due to felling in the early nineteenth century, followed by rapid recovery.[14] The capercaillie, dependent on mature pine, became exterminated as early as the 1780s, and was only introduced again in new maturing forests after a gap of seventy years;[15] the red squirrel, dependent on pine seeds, was almost extinct by the 1840s, and spread again subsequently, aided by reintroductions;[16] the great spotted woodpecker died out at this point, and re-immigrated later.[17]

The revival of these species came with new plantations on a very extensive scale, combined with greater care in regeneration and assisted planting in the traditional pinewoods of the Speyside Valley and elsewhere.[18] Historians have often noted with disapproval the 60,000 trees contracted to be felled by the York Buildings Company when they purchased an interest in Grant lands in the Abernethy Forest in 1728;[19] if indeed they were all felled, this was a substantial loss of mature trees from the already depleted gene bank of the old forests, though one cannot help but notice that the Abernethy forest is still there. Such figures need to be set against, say, the 31 million saplings planted by Sir Francis Grant on his estates up to 1847, though many of these would die or be thinned long before they could reach the stature of the felled forests.[20] Scots pine appears to have been the favourite tree of the Victorian forester, and it was technically impossible to plant the peat hillsides and

---

[14] Smout, MacDonald and Watson, *Native Woodlands*, ch. 8.

[15] J. A. Harvie-Brown, *The Capercaillie in Scotland* (Edinburgh, 1879); Ritchie, *Animal Life*, pp. 269–70. A recent thesis argues that capercaillie extinction was exacerbated by climate change, as well as by human persecution and problems with the habitat in the 1780s: G. B. Stevenson, 'An historical account of the social and economic causes of capercaillie, *Tetrao urogallus*, extinction and reintroduction in Scotland', unpublished University of Stirling Ph.D. thesis, 2007.

[16] J. A. Harvie-Brown, *The History of the Squirrel in Great Britain* (Edinburgh, 1881); Ritchie, *Animal Life*, pp. 290–5.

[17] J. A. Harvie-Brown, 'The great spotted woodpecker (*Picus major, L.*) in Scotland', *Annals of Scottish Natural History*, 1 (1892), pp. 4–17.

[18] Smout, MacDonald and Watson, *Native Woodlands*; David Nairne, 'Notes on Highland woods, ancient and modern', *Transactions of the Gaelic Society of Inverness*, 17 (1891); J. M. Lindsay, 'The use of woodland in Argyllshire and Perthshire between 1650 and 1850', unpublished University of Edinburgh Ph.D. thesis, 1974; J. M. Lindsay, 'Forestry and agriculture in the Scottish Highlands, 1700–1850: a problem in estate management', *Agricultural History Review*, 25 (1977), pp. 23–36; G. A. Dixon, 'Forestry in Strathspey in the 1760s', *Scottish Forestry*, 30 (1976), pp. 38–60.

[19] For example, Ritchie, *Animal Life*, p. 321.

[20] Nairne, 'Notes on Highland woods', p. 199.

flows where modern forestry is so intrusive. One could probably summarise the woodland history of nineteenth century Scotland as a decline of oak and birch after a century of relative stability, ending around 1820, but an increase in pine after a century of further decline ending around 1850. At the end of the day, there was probably about 9 per cent of the land covered in wood in about 1800, and about 5 per cent in 1900.[21]

Fraser Darling further maintained that disafforestation had been followed by the over-exploitation of the land by fire and sheep: the stored nutrients in once-wooded ground had been dispersed by the smoke of muirburn, designed to 'improve' the grazing, and driven off in the wool, meat and bone of the sheep that took it from the vegetation. In his own words, 'two centuries of extractive sheep farming in the Highland hills have reduced a rich natural resource to a state of desolation', and elsewhere he drew analogies with twentieth-century Africa to illustrate the seriousness of what he thought had happened.[22] Modern research has not ruled out this effect, but has not confirmed it either. It is beyond doubt that vegetation is modified in different ways by the grazing pressures of sheep and cattle, though it is less easy to prove experimentally that sheep bring about a deterioration of the biological carrying capacity of the ground. There is evidence that the formation of heather moor of the sort that follows forest removal and grazing by cattle can itself also lead to acidification, and that under some circumstances sheep grazing recycles nutrients more rapidly and may actually increase soil fertility, so that grasslands intensively grazed by sheep may be richer in plant species than less-intensively grazed grassland.[23] If the sheep are taken off, a heavy mat of coarse grass may replace both the former heather and the close-cropped sward, which perhaps happened in many areas when sheep numbers declined in favour of rising deer numbers between the 1880s and the 1930s. What the Victorians were very keen on protecting, of course, was game. The rise of the sporting estate followed on the imposition of English-

---

[21] Ibid. p. 191.

[22] Fraser Darling, 'Ecology of land use in the Highlands and Islands', in D. S. Thomson and I. Grimble, *The Future of the Highlands* (London, 1968), p. 38; the phrase 'wet desert' was perhaps first used by him in *Pelican in the Wilderness* (1956), p. 180, and others similarly came to refer to the Highlands as 'semi-desert': W. H. Pearsall, 'Problems of conservation in the Highlands', *Institute of Biology Journal*, 7 (1960), p. 7; W. J. Eggeling in *Natural Resources in Scotland: Symposium at the Royal Society of Edinburgh* (Scottish Council, Development and Industry, 1961), p. 362.

[23] J. Miles, 'The pedogenic effects of different species and vegetation types and the implications of succession', *Journal of Soil Science*, 36 (1985), pp. 571–80, esp. p. 575. See also R. F. Hunter, 'Conservation and grazing' in 'Land utilization and conservation in the Scottish Highlands', *Institute of Biology Journal*, 7 (1960), pp. 20–2; T. C. Smout, *Nature Contested: Environmental History in Scotland and Northern England since 1600* (Edinburgh, 2000), ch. 5.

style game laws in the latter part of the eighteenth century,[24] and on the knowledge of the wealth of quarry by the outside world in the 1780s and 1790s.[25] By the end of the Napoleonic Wars, the Duke of Gordon was advertising the delights of his shooting leases in the south, and the full-blown sporting estate was on the market before mid-century.[26] With the collapse of mutton and wool prices in the years after 1870, the deer forest and the grouse moor became the prime use of much Highland land and, by the early decades of the twentieth century, it was a matter of hot political contention as to whether deer forests represented the sterilisation of a valuable resource for the amusement of the idle and alien rich or the only practicable use for otherwise almost valueless land. Meanwhile the number of deer increased enormously over what it had been in the eighteenth century and, like sheep, came to represent an ecological threat to woodland regeneration. Adam Watson has drawn attention to eighteenth-century clearances for game preservation on several Cairngorm estates and the subsequent failure, after about 1800, of many of the older Scots pinewoods to regenerate naturally due to browsing pressures from the ever-increasing numbers of red deer.[27]

With the game came gamekeepers, and technology in due course supplied them with deadly weapons against predators, the cartridge-loading shotgun and the steel gin trap.[28] Early game books are full of details of the destruction of birds of prey on a scale which has occasionally tested credulity: in five Aberdeenshire parishes clustering around Braemar, 70 eagles, and 2,520 hawks and kites are said to have been killed between 1776 and 1786;[29] on the Sutherland estates of Langwell and Sandside, 295 adult eagles were destroyed between 1819 and 1826;[30] and on the single estate of Glengarry

---

[24] Adam Watson and Elizabeth Allan, 'Papers relating to game poaching on Deeside, 1766–1832', *Northern Scotland*, 7 (1986), pp. 39–45.

[25] T. Thornton, *A Sporting Tour through the Northern Parts of England, a Great Part of the Highlands of Scotland* (London, 1804) referred to excursions in the 1780s; Robert Heron, *Observations made in a Journey through the Western Counties of Scotland in the autumn of 1792* (Perth, 1793) spoke of the attractions for sportsmen. See S. Nenadic, 'Land, the landed and the relationship with England: literature and perception 1760–1830', in S. J. Connolly, R. A. Houston and R. J. Morris, *Conflict, Identity and Economic Development: Ireland and Scotland, 1600–1939* (Preston, 1995), pp. 148–60.

[26] D. Hart-Davis, *Monarchs of the Glen* (London, 1978); Smout, 'Tours in the Scottish Highlands', pp. 110–11.

[27] A. Watson, 'Eighteenth century deer numbers and pine regeneration near Braemar, Scotland', *Biological Conservation*, 25 (1983), pp. 289–305; A. Watson and E. Allan, 'Depopulation by clearances and non-enforced emigration in the north-east Highlands', *Northern Scotland*, 10 (1990), pp. 31–46. See also Desmond Nethersole-Thompson and Adam Watson, *The Cairngorms* (Perth, 1981), pp. 20–7.

[28] David Allen, *The Naturalist in Britain: a Social History* (London, 1976), pp. 141–3.

[29] Ritchie, *Animal Life*, pp. 128–36.

[30] Ibid. p. 128.

well over a thousand kestrels and buzzards, 275 kites, 98 peregrine falcons, 78 merlins, 63 hen harriers, 63 goshawks, 106 owls, 18 ospreys, 42 eagles and sundry other hawks, were killed in only three years, 1837–40.[31] To take a lowland example, 310 hen harriers were killed on one Ayrshire estate in four years of the nineteenth century.[32] My own view is that there is so much of this evidence from such varied sources that it cannot be disregarded in total, though some details of species identification may be wrong. The destruction of the birds of prey on this scale was itself a major modification of the natural world. Small mammal predators suffered at least as badly. Glengarry estate destroyed some 650 pine martens, wild cats, polecats, badgers and otters in three years,[33] and James Ritchie provided astonishing details of the skins brought to the Dumfries fur market in the nineteenth century: 600 polecat skins a year in the 1830s.[34] The polecat was extinct in Scotland by the end of the nineteenth century.

At least as significant as the fact of their destruction is what the former volume of predators reveals about the volume of prey in the Highlands – voles, mice, hares, small birds and so on.[35] They clearly no longer exist at anything like the densities necessary to support such numbers of predators, presumably because of damage done to their habitat by two centuries of modern land use.

Victorian society seems to have paid little attention to the alteration and destruction occurring in the Highlands. It is true that you may, if you search, happen upon a denunciation of the new, closely planted conifer woods as savage as anything that might be said in the 1980s:

> The whole is one enormous, unbroken, unvaried mess of black . . . it has some of that dreary image, that extinction of form and colour, which Milton felt from blindness . . . the whole wood is a collection of tall, naked poles, with a few ragged boughs near the top . . . below, the soil parched and blasted with the baleful droppings; hardly a plant or a blade of grass; nothing that can give an idea of life or vegetation.

---

[31] Roger Lovegrove, *Silent Fields: the Long Decline of a Nation's Wildlife* (Oxford, 2007), p. 74; Edward C. Ellice, *Place Names of Glengarry and Glenquoich* (London, 1931), p. 19.

[32] W. H. Pearsall, *Mountains and Moorlands* (London, 1971), p. 236. Some of the predators slain in these Victorian holocausts may have been migrants: others will have been drawn in to meet their fates by gaps in the predator territories created by earlier killings. One should not, for example, conclude that one estate could ever have supported 310 resident hen harriers in a four-year period.

[33] Ellice, *Place Names*, p. 19.

[34] Ritchie, *Animal Life*, pp. 165–7.

[35] For enormous eighteenth-century game bags, see Thornton, *A Sporting Tour*.

A man might choose to hang himself there, but would find it hard to discover a side stem to which a rope could be fastened.[36]

More notably, John Ruskin, that son of Perth, drew attention in 1886 to the desecration of an inlet of Loch Katrine associated with Walter Scott in a passage interesting for its early use of the concept 'inheritance' to describe Scottish natural beauty:

> That inlet . . . was in itself an extremely rare thing; I have never myself seen the like of it in lake shores. A winding recess of deep water, without any entering stream to account for it – possible only, I imagine, among rocks of the quite abnormal confusion of the Trossachs, and beside the natural sweetness and wonder of it, made sacred by the most beautiful poem that Scotland ever sang by her stream-sides. And all that the nineteenth century conceived of wise and right to do with this piece of mountain inheritance was to thrust the nose of a steamer into it, plank its blaeberries over with a platform, and drive the populace headlong past it as fast as they can scuffle.[37]

On the other hand, there were no campaigns in Scotland analogous to the defence of the Lake District in England, which was an element in leading to the foundation of the National Trust in 1894.[38] Then there was no attempt for almost forty years to interest the National Trust in properties in Scotland (though it quickly acquired them in Wales and Ireland), and no immediate effort to create an analogous Scottish body. There was no Scottish nature writer working to raise a green consciousness analogous to Richard Jefferies and W. H. Hudson in England or, perhaps more remarkably, to the great John Muir, born a Scot, in western America. Similarly, there appears to be no analogy within Scotland to the Scottish surgeons and missionaries who even before the middle of the nineteenth century had provided a precocious and trenchant critique of environmental damage in India and South Africa, as studied by Richard Grove.[39]

Again, when the Royal Society for the Protection of Birds (RSPB) was founded in 1889 it acquired some early Scottish members and, in 1893, a

---

[36] Quoted in Veitch, *Feeling for Nature*, vol. 2, pp. 152–3.

[37] John Ruskin, *Praeterita* (London, 1885–6), vol. 1, pp. 411–12.

[38] John Sheail, *Nature in Trust: the History of Nature Conservation in Britain* (Glasgow, 1976), p. 59.

[39] R. H. Grove, *Green Imperialism: Colonial Expansion, Tropical Island Edens and the Origins of Environmentalism, 1600–1860* (Cambridge, 1995), ch. 8; R. H. Grove, 'Scottish missionaries, evangelical discourses and the origins of conservation thinking in Southern Africa 1820–1900', *Journal of Southern African Studies*, 15 (1989), pp. 163–87.

Scottish vice-president, but remained firmly rooted in England.[40] Early bird protection legislation was not focused on the Highlands even when it had a Scottish application. The first area nominated a bird sanctuary by a Scottish local authority under the 1894 act was in the Lowlands at Tentsmuir in Fife; the earliest species to attract individual attention in statutes were seabirds and the St Kilda wren.[41] Why this should have been the case is hard to say, but it has been a persistent characteristic that Scotland in general, and the Highlands in particular, have lagged behind the rest of Britain in popular interest in formal conservation of nature and landscape.[42]

Nevertheless, there were two very significant developments for the future in Scotland before the nineteenth century drew to a close. One was the rise of a movement to secure public rights of access. The Scottish Rights of Way Society was formed in Edinburgh in 1843 primarily to maintain footpaths near the capital from the encroachment of the landed who wished to close them: the civic leaders observed that it was a cheaper way to maintain the health of the populace than by following the expensive recommendation of Edwin Chadwick on sanitary provision.[43] Almost at once they were embroiled, with the support of Perth Town Council and others, in a famous dispute about access down an old drove road through Glen Tilt between an Edinburgh professor leading an excursion of science students and one of Scotland's greatest sporting landlords and friend of the Royal Family, the Duke of Atholl. The upshot, after further incidents, a cartoon hostile to the Duke in *Punch* and a prolonged series of lawsuits, was that such rights of way became firmly established.[44]

The society fought later on many occasions, for example to secure the path through Glen Doll and the Larig Ghru, always in the teeth of lairdly opposi-

[40] Tony Samstag, *For Love of Birds: the Story of the Royal Society for the Protection of Birds 1889–1988* (RSPB, 1988); Sheail, *Nature in Trust*.

[41] J. A. Harvie-Brown, *A Fauna of the Tay Basin and Strathmore* (Edinburgh, 1906), pp. xli–xliii; J. Morton Boyd, 'Nature conservation', in *Proceedings (Section B) of the Royal Society of Edinburgh*, 84 (1983), p. 296.

[42] Indications of this included in the 1980s much lower membership of the RSPB in Scotland than England, the low membership of the Scottish Wildlife Trust compared to English County Naturalist Trusts and its weakness in the Highlands, the low membership of the Ramblers Association in Scotland, the lower vote for the Green Party in the European elections of 1989, and a low uptake of NCC grants for school conservation projects in Scotland compared to England. On the other hand, the 1990 Regional Council Elections produced Scotland's first Green Party councillor (in Nairn) and the party scored unusually well in votes in Highland Region.

[43] National Archives of Scotland (NAS): G.D. 335, Records of the Scottish Rights of Way Society.

[44] Tom Stephenson, *Forbidden Land: the Struggle for Access to Mountain and Moorland* (Manchester, 1989), pp. 120–3; NAS: G.D. 335.

tion, often with surreptitious support from the local population.[45] The cause of access, in England and Scotland alike, became a cause of the radical Liberals in their campaign against the wicked privileges of the landed classes, exemplified from 1884 by James Bryce's many repeated attempts to bring in an Access to the Mountains Bill that would have opened the hills to everyone and hopefully made life hell for the aristocratic deer stalker and the grouse shooter.[46]

Aside from the political aggression, it was a sign of the times that there was a great increase in the recreational use of the mountains by middle-class town dwellers. The first number of the *Scottish Mountaineering Club Journal* appeared in 1890, and the president of the club declared that 'the love of scenery and of the hills is implanted in the heart of every Scot as part of his very birthright', and on the tops 'we seem to breathe something else than air, and . . . look down on every side upon a scene untainted by work of man, just as it came fresh from the Creator's hand'.[47] This was good mystical green stuff, reaching back to the world of Wordsworth and forward to the Friends of the Earth. The Scottish Ski Club was founded in the new century, by fourteen gentlemen from Edinburgh in 1907. They gave away skis to shepherds and postmen, but the tone of their journal was much more macho:

I glory in the victory over self and Nature . . . the greatest of all joys of ski-ing is the sense of limitless speed, the unfettered rush through the air at breakneck speed . . . Man is alone, gloriously alone against the inanimate universe . . . He alone is Man, for whose enjoyment and use Nature exists.[48]

Thus encapsulated are two aesthetic attitudes to the hills, as temple and as obstacle course, which exist to this day and can be seen as opposition and support for the development of the Northern Corries of Cairngorm in 1990.

The other significant development of the turn of the nineteenth and twentieth centuries was the Scottish birth of ecology as a scientific discipline.[49] I am not trying to say that the Scots invented ecology: it had European origins in the late nineteenth century. But there was a special Scottish context for the study of ecology. It was associated particularly with the teaching of zoology

[45] NAS: G.D. 335.
[46] Ernest A. Baker, *The Highlands with Rope and Rucksack* (London, 1923), p. 22; Sheail, *Nature in Trust*, p. 69; Stephenson, *Forbidden Land*, pp. 131–42.
[47] *Scottish Mountaineering Club Journal*, 1 (1890), pp. 1–3.
[48] *Scottish Ski Club Magazine*, 1 (1913), p. 207.
[49] C. H. Gimingham, D. H. N. Spence and A. Watson, 'Ecology', in *Proceedings (Section B) of the Royal Society of Edinburgh*, 84 (1983), pp. 86–8; John Sheail, *Seventy-five Years in Ecology: the British Ecological Society* (Oxford, 1987).

and botany as 'Vitalistic science' at the University College of Dundee, under Patrick Geddes, Professor of Botany, and D'Arcy Thompson, author of *Growth and Form*, marine biologist, classicist and *savant extraordinaire*.[50] It also involved very talented amateur natural historians, of whom J. A. Harvie-Brown, the mastermind of the various *Vertebrate Fauna* of Scottish natural regions, was the most remarkable,[51] and John Hood, the metal-turner of Dundee who discovered 65 of the 400 rotifers known to science in his day, is the most forgotten.[52] It would take a paper in itself to do justice to the figures concerned.[53] To two of Geddes' students, Robert Smith and his brother W. G. Smith, we owe what has been called 'the first attempt at a scientific description of British vegetation', using mapping techniques introduced at Montpellier by Charles Flahault, with whom Geddes himself was in close contact.[54] Robert's untimely death from appendicitis at the age of 26 in 1900, and his brother's move into agricultural science after 1908 terminated the work, 'and thereafter the subject made virtually no progress for thirty to forty years'.[55] Another in the same mould was W. R. Collinge, student at St Andrews in the 1880s, founding editor of the *Journal of Economic Biology* by 1906, and university assistant lecturer in D'Arcy Thompson's St Andrews by 1917, when he started an organisation called the Wild Bird Investigation Society, which sounds like an early forerunner of the British Trust for Ornithology. Within five years, he had left to become Curator of the Yorkshire Museum.[56] Finally, there was James Ritchie, successively at the University of Aberdeen, the Royal Scottish Museum and the University of Edinburgh, and thus one of the few noted zoologists to remain in Scotland through the first half of the twentieth century. In 1920, he published an important book, its title demonstrating its originality, *The*

---

[50] Allen, *Naturalist in Britain*, p. 242; Patrick Geddes and D'Arcy Thompson were both better at firing people with ideas than teaching them the nitty-gritty of biology; it was said of D'Arcy Thompson later at St Andrews that he imbued his students with knowledge of 'the beauty of the whole animal kingdom', but omitted to teach them the anatomy of the frog. R. D'Arcy Thompson, *D'Arcy Wentworth Thompson: the Scholar Naturalist* (London, 1958), p. 167.

[51] The Harvie-Brown papers are deposited in the Royal Museum of Scotland, Chambers Street, Edinburgh.

[52] A. D. Peacock, 'John Hood of Dundee: working-man naturalist', *Scots Magazine*, 31 (1939).

[53] One additional figure was Norman Kinnear, who did much to encourage the study of natural history and ecology in India, and later followed a distinguished career at the Natural History Museum in South Kensington.

[54] J. H. Burnet (ed.), *The Vegetation of Scotland* (Edinburgh, 1964), pp. 1–4; Sheail, *British Ecological Society*, pp. 6–10.

[55] Burnet, *Vegetation*, p. 1.

[56] He was also involved in the early bird protection movement in Britain and Scotland: W. E. Collinge, 'The necessity of state action for the protection of wild birds', *Avicultural Magazine*, 10 (1919), pp. 123ff.

Influence of Man on Animal Life in Scotland, at least thirty years ahead of its day in scope and subject matter. It had no immediate imitators on either side of the Border.[57]

It was, I believe, a characteristic of much early Scottish ecology not only that it was very innovative about biological mapping (in which Harvie-Brown with the help of the cartographic firm Bartholomew played a role, as well as the Smith brothers), but it was remarkable in seeing man himself as a dynamic prime actor among other animals. Patrick Geddes himself was deeply, indeed primarily, interested in man in his environmental setting. Harvie-Brown's publications on the capercaillie, the red squirrel and the great spotted woodpecker were forerunners of the kind of historical ecology we are only now rediscovering.[58] Collinge's interests in 'economic biology' and 'economic ornithology' made him a pioneer investigator of the impact of animal life on man: his paper on the food of the little owl was a model for other scholars.[59] Ritchie's interests are self-evident.

The predominant search in England, influenced by the circle round the brilliant, disdainful and rigorous A. G. Tansley, was to find objects of ecological study where the impact of man on nature could be stopped or controlled in nature reserves, and the study of the resulting 'anthropogenic climax' pursued with hard, quantifiable biology. Tansley had decisively parted company with Frederic Clements and the dominant American school of ecology, for whom only a wilderness climax unaffected by man was worthy of study. He respected human history as part of nature, but believed once an ideal state had been achieved as a by-product of human activity, such as agriculture turning scrub into herb-rich grassland, that state should be frozen in time.[60] British ecology in the 1940s and 1950s was an influential and crusading discipline, Tansley joining Julian Huxley, Max Nicolson and their colleagues in achieving the establishment of the Nature Conservancy as an agency to protect British wildlife. But at the same time it was in danger of becoming an arcane 'pure' science, almost antiquarian

---

[57] Dudley Stamp was to describe it in the 1960s as a work 'in advance of his time: it includes much which is of the highest value today', *Nature Conservation in Britain* (London, 1962). The next work somewhat along these lines was E. M. Nicholson, *Birds and Men: the Bird Life of British Towns, Villages, Gardens and Farmland* (London, 1951).

[58] He was also the first to attempt a thorough census of a common colonial land bird through co-operative research: by writing to the Caithness landowners to ascertain the status of the rook in 1875; Allen, *Naturalist in Britain*, p. 216.

[59] W. E. Collinge, *The Food of Some British Wild Birds: a Study in Economic Ornithology* (York, 1924); Sheail, *Nature in Trust*, p. 51.

[60] Tansley is said to have been the only man ever to have accepted a Chair in Oxford while continuing to live in Granchester. For his intellectual importance see Sheail, *British Ecological Society*, but this judgement on his legacy is my own.

and concerned with the status quo, unrelated to the social sciences and so irrelevant to society. The Scottish vision was submerged, primarily, I think, because Patrick Geddes switched his maverick genius from ecology to town planning for the rest of his life, and because the Scottish universities in the interwar years offered so few careers for the talented.

The consequent lack of interest by British ecology in general in the dynamic character of man's impact on nature has been described by one of the leading figures in Nature Conservancy science in the 1960s and 1970s, Dr Norman Moore, in these terms:

> Up to the 1950s, human impacts on habitat and species were deliberately or subconsciously excluded from field studies. I suspect the reasons were complicated – a natural desire by biologists to simplify by excluding what seemed irrelevant, a snobbish distaste for the obvious and the practical by the intellectual, and an increasing loss of day-to-day experience of farming practice among most naturalists and biologists as Britain became more and more industrialised.[61]

This was a completely English perspective. Only an ecological science that had turned its back on Geddes, Harvie-Brown and Collinge, and never regarded James Ritchie's great book, let alone considered the work of Frank Fraser Darling from the late 1930s onwards, could have fenced itself into such an untenable corner.

In the interwar years, the public view of the Scottish Highlands was dominated by two developments that ultimately turned out to be on a collision course. The first was the idea that forestry and light industry might stem the continuing, and now seemingly catastrophic, decline of the Highland rural economy. Both sheep farming and crofting were at a very low ebb; the herring markets in Eastern Europe had evaporated with the war and the Russian Revolution; industrial depression, as well as the decline of employment in fish-gutting, knocked the bottom out of the migrant labour market where young Highlanders had participated through the previous century. The result was the most rapid outflow of population ever recorded, despite the Board of Agriculture's laborious and belated attempts to return the land to the people by creating crofts out of former grazing land for anyone who wanted them.[62] Even the sporting estate suffered, with the declining fortunes of the landed classes, changing hands rapidly and at comparatively low prices.

---

[61] N. W. Moore, *The Bird of Time: the Science and Politics of Nature Conservation* (Cambridge, 1987), p. 44.

[62] Leah Leneman, *Fit for Heroes? Land Settlement in Scotland after World War I* (Aberdeen, 1989).

It still seemed to some, however, especially landowners, that the deer forest continued to offer the best of a bad lot of options for viable land use in the Highlands. To others, especially on the left, this conclusion was defeatist and obscene, bearing in mind how many people the hills and glens had supported less than a century before. There were great hopes of the newly constituted Forestry Commission, set up in 1919 to increase the nation's wood supplies in the event of another war. The first plantations were in by 1921, but its ability to buy land was restricted by financial stringency and, until after the Second World War, its technical impact was still limited by the inability to plough into the peaty hillsides in order to grow upland forests.[63]

Others pinned their hopes on light industries powered by hydroelectric schemes. The first schemes, opened by the British Aluminium Company at Foyers in 1896 and Kinlochleven in 1909, were widely recognised as environmental, even social, disasters. Thus said David Kirkwood, Labour MP, in a debate reported in Hansard in 1941:

See what this dastardly crowd have done to a section of the Highlands of Scotland. Go and see the housing conditions ... see the wooden shanties two stories high. If you go into the side streets, there are back-to-back houses ... two great iron chimneys, two great heaps of dross, mud flung all over the place, pipes coming down the beautiful hillside which is all denuded of plants, trees and every kind of vegetation ... Do you mean to tell me that the individuals behind this thing are more interested in my native land than the people who live there? It is just a lot of nonsense.[64]

The next schemes, by the Grampian Power Company for Rannoch and Tummel were authorised in 1922 and opened in 1930 and 1933 respectively, but were much criticised both for their environmental damage and for their failure to sell power to local industry. The plans of the Caledonian Power Company were to capture the headwaters of east-flowing rivers for the benefit of schemes in the west, and led to civil war between Highland local authorities; it would have had a disastrous impact on Loch Ness and many other glens. No fewer than six Highland hydro schemes were promoted in Parliament between 1929 and 1941, and all were rejected.[65] Most were

---

[63] M. V. Edwards, 'Peat afforestation', in 'Land utilization and conservation in the Scottish Highlands', *Institute of Biology Journal*, 7 (1960), pp. 16–19.

[64] Hansard, 374 (1940–1), p. 255.

[65] Peter L. Payne, *The Hydro: a Study of the Development of the Major Hydro-Electric Schemes undertaken by the North of Scotland Hydro-Electric Board* (Aberdeen, 1989); K. J. Lea, 'Hydro-electric power developments and the landscape in the Highlands of Scotland',

overwhelmingly supported by a majority of Scottish MPs, but destroyed by the English majority in Westminster on grounds that ranged from the naked self-interest of landowners defending sporting interests, to genuine passion about the environmental impact. In the debate on the Grampian Electricity Supply Order in 1941, the Secretary of State for Scotland, Tom Johnston, said 'not since the English prayer book issue have I witnessed such division of opinion in various parties of the house'.[66] Noel Baker, the Labour MP, spoke of the proposals to flood Glen Affric and Glen Cannich as 'not merely a matter for the people of the Highlands. This is a matter for the whole British nation ... This is not a question of minor or transient aims. The Highlands are the spiritual heritage of the whole people which it is the duty of Parliament to preserve'.[67] The Conservative member for Twickenham, Keeling, was on the same side:

> I know that a great many people do not care for nature at all, and would boil down the last nightingale if they could get a farthing's worth of glue out of his bones. But to those people I will put two practical questions. The first question is, Do they not realise that beautiful scenery has commercial value, that it attracts an enormous number of tourists, and that tourist traffic is one of the main assets of the Highlands? ... The second question I put to those who do not care for beauty is this. Do they not realise that there are a vast number of men and women who do care passionately? ... every time you injure a glen such as Glen Affric you are injuring very many people who depend upon beautiful scenery to restore their nerves and to revive their minds and souls.[68]

These two arguments, the economic one about the tourist trade and the green or Wordsworthian one about spiritual renewal, were to be repeated in every serious debate on Highland conservation in the next fifty years. It is noteworthy that nothing at all was said in this or any other of the Parliamentary debates of the time about ecology or wildlife: the 'heritage' is still entirely scenic.

The second development which was to be of the greatest long-term significance to the history of conservation in Scotland was the emergence of

---

Scottish Geographic Magazine, 84 (1968), pp. 239–55; North of Scotland Hydro-electric Board, *Power from the Glens* (Edinburgh, 1976), p. 2; J. Sheail, *Rural Conservation in Inter-war Britain* (Oxford, 1981), pp. 39–46.

[66] Hansard, 374 (1940–1), p. 255.
[67] Ibid. p. 240.
[68] Ibid. p. 207.

a movement to promote national parks on the model of the United States, where the world's first national park had been established at Yellowstone in 1872, and on the model of Canada. The first proposal appeared as early as 1904, when Charles Stewart contributed an essay to *The Nineteenth Century and After* proposing the 'preservation in its wild state of a large tract of country possessing a natural beauty and grandeur in high degree'. He thought Jura or Rum might be bought by the state, or one or more of the great Highland deer forests, and he saw benefits for public recreation, for science and for the preservation of wildlife. He even foresaw that the creation of a forestry commission – of which he heartily approved – could create what he called 'an additional and urgent reason for preserving those wild animals whose extinction may be threatened by the substitution of extensive forestry for deer forests and grouse moors'.[69]

All this was visionary; it fell on stone-deaf ears at the time, but in 1928 Lord Bledisloe, Parliamentary Secretary to the Ministry of Agriculture, returned from a private visit to Canada and the USA fired with enthusiasm for what he had seen and anxious to see the creation of British national parks in the Conservative Party election manifesto.[70] At almost exactly the same time, in the autumn of 1928, the *Scots Magazine* began to orchestrate a campaign for the purchase of the Cairngorms as a national park.[71] The idea owed a great deal to J. W. Gregory, Professor of Geology at Glasgow and past-president of the Scottish Ski Club. It had the backing of rambling interests and of the recently formed Association for the Preservation of Rural Scotland, and claimed wide political support from Tories like Walter Elliot and John Buchan, from Labour members like Tom Johnston ('the Park must be for all the people, and not for a few plutocrats only': no Gleneagles prices), from Edwin Scrymgeour the temperance reformer and Erskine of Marr the Nationalist ('this Caledonian forest should be a National Sanctuary – a sort of serpentless Eden').[72] Ramsay Macdonald gave his 'whole-hearted support to the idea . . . the flora and fauna must be considered in any proposed plan.

---

[69] Charles Stewart, *In the Evening* (London, 1909), pp. 51–64.

[70] John Sheail, 'The concept of national parks in Great Britain, 1900–1950', *Transactions of the Institute of British Geographers*, 65 (1975), pp. 41–56; Sheail, *Rural Conservation*, p. 115.

[71] *Scots Magazine*, 10 (1928–9) – especially, E. A. Baker, 'The Cairngorms as a National Park'; Alan Graeme, 'Hills of Home'; J. W. Gregory, 'The claim for a National Park'; A. I. McConnochie, 'The national nature of the Cairngorms'. *The Cairngorm Club Journal* (January 1930) confirmed that the credit for launching the idea of the Cairngorms as a National Park rested with the *Scots Magazine*, leading to enthusiasm in other papers, especially the *Glasgow Herald*. In June 1929, the Association for the Preservation of Rural Scotland called a conference of out-door organisations, and a committee styled the Scottish Forest Reserve Committee was set up to promote the idea nationally.

[72] *Scots Magazine*, 10 (1928–9), 'Scottish MPs and the National Park', p. 337, 'The claim for a National Park', pp. 416–19.

This is a point of the very highest importance'; he had reservations about the development of winter sports that might leave 'the Cairngorms spotted with miscellaneous erections'.[73]

It was of course Ramsay Macdonald who subsequently, in 1929, appointed the Addison Committee to investigate the possibility of British national parks. This proved a false start amid the crises and distractions of the following decade. It was at least a forerunner to the Dower Report and other government commissions and committees leading to the establishment in 1949 of the Nature Conservancy and the National Parks Commission, the latter with powers in England and Wales alone. This is a story well told in detail by John Sheail and others, and need not be repeated here.[74] But in Scotland the immediate impact of the *Scots Magazine* campaign was very different. It was at once clear that the government was not going to buy Cairngorm for the nation. Would private interests do so? The National Trust in the south was approached, a fact which may well have acted as something of a catalyst for the formation of a National Trust for Scotland (NTS) in 1931; presumably the reproach was too much to bear.[75] Cairngorm was not on the market, but before the Second World War the NTS had secured large areas in Glencoe and Ross-shire for public preservation and enjoyment. Their role as a conservation body for the countryside, as opposed to historic houses, was relatively more prominent than it is today.[76] The other body enthusiastically to enter the ring was the Forestry Commission, which offered the unplantable parts of their Glenmore holdings in Cairngorm for public access, and then more decisively launched the first National Forest Park at Ardgarten in Argyll in 1936, though the Treasury strongly objected to 'converting the Commission appointed to re-afforest Great Britain into an agency for promoting "hiking"'.[77]

Indeed, almost all the emphasis in the Scottish national parks movement and its allies at this stage was on opening up the countryside for public recreation. Little need was seen for specific wildlife conservation, since it was assumed at first that there would be little or no conflict between the nature

---

[73] *Scots Magazine*, 10 (1928–9), J. R. Macdonald, 'Why I support the Cairngorm National Park Scheme', pp. 251–2.

[74] Sheail, *Nature in Trust*; Gordon Cherry, *National Parks and Recreation in the Countryside, Environmental Planning 1939–1969* (HMSO, 1975), vol. 2.

[75] *Scots Magazine*, 13 (1929), S. E. Hamer, 'The National Trust in England'; 14 (1930–1), Alan Graeme, 'The National Trust'; Sheail, *Nature in Trust*, p. 60.

[76] Yet as late as 1969, Fraser Darling declared 'I would say the Scottish Trust now leads the world in the wholeness of its approach to environmental management': *Wilderness and Plenty*, p. 67. See also Morton Boyd, 'Nature conservation', p. 297.

[77] Sheail, *Rural Conservation*, pp. 172–86; *Scots Magazine*, 23 (1935), 'Scotland month by month'.

reserve and the national park idea. With the foundation of the Scottish Youth Hostel Association in 1931, there was a very large increase in rambling in the Highlands; its founders saw it as patriotic, helping to 'build up a stronger and more generally useful generation for the service of the state' and mixing together 'young people of very diverse station and calling in life in the common companionship of tired limbs round the hostel fire in the evening'.[78] There was no comparable explosion of popular interest at this stage in bird watching or natural history. Ecological science in the Scottish universities was similarly going through a lean patch. The most imaginative effort in these years came from Dudley Stamp in England proposing a Land Utilization Survey, very much in the Patrick Geddes tradition. The Department of Agriculture for Scotland was no friend to surveys and the exercise became more limited than it would otherwise have been.[79]

It was to this scene that there arrived in the 1930s a young English biologist with an early training in agriculture and a doctorate in genetics of the blackfaced sheep from Edinburgh University, Frank Fraser Darling, a man of compulsive inner drive, considerable literary gifts, and a visionary outlook on man and nature, that blended science and the romantic tradition. For him, 'science and art are not far from one another'.[80] He was also a man of charisma and little tact. From 1935, when he took up a Leverhulme Fellowship studying red deer in Dundonnell, until 1959, when he left Scotland to become Vice President of the Conservation Foundation in Washington, DC for thirteen years, he was the most compelling personality in conservation in Scotland. The vision that he had was that man was inescapably part of the natural ecosystem and that the whole of nature was in every way, and all the time, exposed to man. Such ideas were not exactly original to Fraser Darling – he owed a considerable debt to American writers such as Aldo Leopold and William Vogt – but he was certainly the first to translate them to a Scottish context.[81] It is a measure of how far we have, at least superficially, come since mid-century that such thoughts may seem to us almost banal today, but few others in any position of influence in Britain in the 1940s and 1950s shared them. The common-sense view at that time was still that there was 'man', on the one hand, and 'the natural world' on the other. The two might impinge on one another, mainly in the twentieth century by man invading the natural world, and it was therefore

[78] *Scots Magazine*, 17 (1932), Alexander Gray, 'The Scottish Youth Hostels Movement', pp. 81–6.
[79] Sheail, *Rural Conservation*, p. 163.
[80] Fraser Darling, *Island Farm*, p. 12. An excellent biography for the Scottish parts of his career is J. Morton Boyd, *Fraser Darling's Islands* (Edinburgh, 1986).
[81] See Fraser Darling, *Pelican in the Wilderness*, p. 21.

proper for man to lay aside certain small pieces of the earth as reserves to allow the natural world to proceed unhindered within them. It was the business of ecological science to study the natural world in such reserves and the business of such reserves to provide the materials for science. Hence the terminology, still with us, of NNRs and SSSI. For Fraser Darling, it was as though the whole globe was a Site of Special Scientific Interest, and the main business of ecology should be to arouse humanity to appreciate its precarious and responsible position upon it.

It is important to appreciate that in many ways Fraser Darling did not enjoy the respect of his scientific peers in the universities and in the British Ecological Society to the degree that might have been expected. Though he spent years of his life in remote places – the Summer Isles, North Rona, Dundonell – living in the close proximity of the sea birds, seals and deer that he was studying (most unusual in Britain then though commonplace now), his science was regarded as good at the intuitive level, but also as soft, non-quantitative, on too large a canvas, not wholly free from the fatal taint of anthropomorphism.[82] Books like *A Herd of Red Deer* (1937) and *A Naturalist on Rona* (1939) were published when the study of animal ecology was emerging as a distinctive branch of the subject (plants had dominated in Britain before the work of Charles Elton in the 1930s),[83] but they were also alive with descriptive passages about the scenic context that the scientific world of Tansley and his colleagues found out of place.

His most original work, *West Highland Survey: an Essay in Human Ecology*, emerged from a proposal he made to Tom Johnston in 1943 to study the underlying causes of depopulation and economic decline in the West Highlands in an ecological context. Its message was that there could be no cure unless the misuse of the land over the two previous centuries was reversed by a modified ratio of sheep to cattle (fewer sheep, more cattle) and by reafforestation. The Department of Agriculture for Scotland, beginning to operate in a world of subsidies that looked too good to refuse merely on ecological grounds, did not like it and publication was delayed until 1955.[84] Scientific ecologists in Britain ignored it because it was about man, not an animal they studied. And what was probably his most popular book, *The Highlands and Islands* for the New Naturalist Series, was comprehensively panned in review in 1948 by Wynne-Edwards, the newly appointed Regius Professor of Natural History at Aberdeen, for its 'surprising number of half-truths and errors'.[85] Fraser Darling found much more recognition for

[82] Morton Boyd, *Fraser Darling's Islands*, pp. 103–9.
[83] Sheail, *British Ecological Society*, p. 83.
[84] Morton Boyd, *Fraser Darling's Islands*, pp. 202–9.
[85] Ibid. pp. 223–4.

his work overseas: in the United States, where *West Highland Survey* had been read and admired and where he eventually attained the highest professional honours, and in Africa, where the dependency between man and the environment was so much more obvious than in Scotland. In the very different climate of Britain in 1969, he was invited to give a memorable series of Reith lectures on the environment[86] and in 1970 he was knighted for his services to conservation.

It is, however, the argument of this lecture that recognition came twenty years too late – recognition both for Fraser Darling as a person and, much more importantly, for the attitude towards man-in-nature that Fraser Darling held. Scotland in 1950 did not have a very highly developed sense of environmental consciousness. It was left out of the National Parks legislation, mainly because of the anxiety of the Scottish Office and of Tom Johnston not to allow anything that might obstruct the development of the glens by the new Hydro Board, but also because of the implacable opposition of many landowners and because the NTS and the Forestry Commission believed they unaided could meet the need for scenic conservation and public access.[87]

However, the Nature Conservancy was extended to Scotland in the postwar legislation. The question arose as to whether there should be separate bodies for England and Scotland. The balance was tipped by a strongly worded memorandum from Fraser Darling in favour of one national agency with equal representation for England and Scotland: 'there is no good biological reason for splitting Great Britain into two parts for the purposes of conservation of wild life and research . . . wild life does not observe political frontiers'. His prime objection to separation was a fear that a Scottish wildlife commission would become a prisoner of the Department of Agriculture for Scotland.[88]

There is little doubt that Fraser Darling expected to be asked to head the Scottish division of a British Nature Conservancy. In the event, the Secretary of State for Scotland chose another man, John Berry, gifted biologist, law graduate, ex-intelligence officer and environmental adviser to the Hydro

---

[86] Published the following year as *Wilderness and Plenty*.
[87] Cherry, *National Parks*, pp. 70–80, 141–52; Sheail, *Nature in Trust*, pp. 205ff.; J. Berry, *pers comm*. The attitude of the Forestry Commission to National Parks is clear from NAS: F.C. 9/1. 'Commission meeting 19 January 1944 to discuss the Dower Report', when J. Bannerman, the Scottish representative, said that the effect of the recommendations would be 'to sterilise large sections of the countryside'; he was in due course appointed to the Scottish National Parks Commission. For the attitude of the National Trust for Scotland, see successive *Newsletters*, e.g. 2 (1949), 14 (1956), 18 (1958).
[88] NAS: F.C. 9/3. 'Scottish Wild Life Conservation Committee: Reservation by Dr Darling, 1 March 1947'.

Board, creating the position of Director of Nature Conservation in Scotland; when the Nature Conservancy was set up shortly afterwards he headed the Edinburgh office with distinction for the next eighteen years. Tom Johnston told Berry he had been chosen because the Secretary of State could keep an eye on him, but that Fraser Darling was too wayward a man to handle.[89] The truth was probably more complex. Fraser Darling was in most respects (in terms of intellectual and practical experience) the obvious choice. He was, however, utterly out of tune with influential people in St Andrews House, and had the handicap in Scotland of being an Englishman and an outspoken critic of Scottish landowners and their gamekeepers. Berry, on the other hand, in the official eye, had shown his paces for this particular job as wartime press censor, where he had combined tact with independence and a certain firmness in standing up to authority, including on occasion to that of Tom Johnston himself. It was also an advantage that he happened to be a member of the Scottish Landowners Federation.

Meanwhile, in London the position of Director General fell to Cyril Diver, an accepted amateur ecologist and the civil servant previously responsible for drafting the legislation that set up the Nature Conservancy. He had limited knowledge of or sympathy towards Scottish peculiarities. Under Berry in Scotland, a policy nevertheless emerged of securing much larger National Nature Reserves than in England, beginning with the purchase of a whole mountain, Beinn Eighe, in 1951, and continuing by management agreement as well as by purchase, perhaps most notably in the establishment of the Cairngorm NNR of 64,000 acres in 1954.[90] Secretaries of State for Scotland, both Labour and Conservative, came to support this policy, despite political criticism.

The emphasis on reserves, dictated as it was by Act of Parliament, was entirely informed by the conventional English philosophy that man and development have one sphere (of course by far the biggest), nature and conservation another (a haven, perhaps, for sportsmen and scientists). There was little place in the legislation for the kind of comprehensive rural planning, the delicate consideration of man's place in the environment and appropriate overall strategies for land use, that a Nature Conservancy would undoubtedly have edged towards had Fraser Darling's views prevailed. 'Wild-life

---

[89] Dr Berry, *pers. comm.*

[90] Dudley Stamp, *Nature Conservation*, pp. 169–78; E. M. Nicholson, *Britain's Nature Reserves* (London, 1957), pp. 121–35. The Ritchie Committee (of which Fraser Darling was an important member) had envisaged nature reserves of varying scale, but the way policy developed in Scotland and England showed interesting contrasts. Scotland before 1968 had a much greater interest in creating large national nature reserves but neglected (relatively speaking) the creation of SSSIs; in England, national nature reserves were small but SSSIs were numerous.

conservation would extend over the whole country, and not only over Parks areas or those additional areas to be scheduled as Nature Reserves' he had said in his memorandum of 1947.[91] Berry saw as clearly as Fraser Darling the need to educate the public in the wider issues of nature conservation but, for his use of the press, was criticised by Diver as a 'vulgar publicist'.[92] When Max Nicholson took over as Director General in 1952 he was vastly more supportive of Berry, and of his deputy and successor Joe Eggeling, but all were heavily constrained by the legislation and what had now become the expectations of the Conservancy's correct sphere of operations.

In the 1960s, the extent of global pollution by pesticides was revealed to the public by Rachel Carson's *Silent Spring* in America, and in Britain scientifically confirmed by Nature Conservancy work on the peregrine falcon and golden eagle.[93] Only at this point did an appreciation of the mutual dependence of man and nature even begin to be felt in the corridors of power, and progress since then has, in the judgement of many, been too little and too late.[94] Patrick Geddes as long ago as 1899 had been instrumental in adding 'nature study' to the curriculum of government schools in Scotland (England followed in 1900),[95] but weaknesses in environmental education were identified as the Achilles heel in any programme for conservation half a century later. Fraser Darling said in 1946 'the first thing – the important thing ... is largely the teaching of an attitude of mind to wild life', but neither the Nature Conservancy nor the Scottish Education Department ever subsequently made it a priority. The kind of complaints that were heard then – 'natural history is on the curriculum in every school but I frankly admit it is not stressed ... there is a lack of enthusiasm and interest in nature study amongst teachers' – can again be repeated with only minor changes of emphasis and language forty years later.[96]

In my concluding section, I would like to draw some threads together. It is possible to classify attitudes towards land use in many different ways. Perhaps six main types of attitude have dominated Scottish experience in

---

[91] NAS: F.C. 9/3. 'Reservation by Dr Darling'.

[92] Dr Berry, *pers. comm.*

[93] Moore, *Bird of Time*, especially ch. 16; John Sheail, *Pesticides and Nature Conservation: the British Experience 1950–1975* (Oxford, 1975). Carson's book was more polemical than scientific, and Fraser Darling said of it 'although I would not have liked to write the book myself, I was very glad it had been written': *Wilderness and Plenty*, p. 45.

[94] A statement of government intentions in Scotland in 1990 declares a wish: '(1) To maintain essential ecological processes and life-support systems. (2) To preserve genetic diversity. (3) To ensure the sustainable utilisation of species and ecosystems': *Scotland's Natural Heritage: the Way Ahead* (Scottish Development Department, August 1990), p. 3.

[95] Allen, *Naturalist in Britain*, p. 203.

[96] NAS: F.C. 9/3. 'Note of a discussion held in St Andrew's House on Wednesday 29 May 1946 with representatives of the SED'.

the last 250 years, and they fall into two broad categories. There are three types that I would group together as traditional, and three I would group together as post-romantic. Of the traditional types, the first is to regard the land as a resource from which to make a living by farming, forestry and commercial fishing; and the second to regard the land as a resource for the private, aristocratic pursuits of hunting, shooting and sport fishing. These two attitudes have existed since time immemorial; they were the only attitudes known before the eighteenth century, and completely dominated land use, law and public policy in the Highlands throughout the eighteenth and nineteenth centuries, though in varying proportions and with very varying consequences for those who lived there, depending on precisely what was involved. Small-scale crofting, associated with potato husbandry and kelping on the islands, which dominated 1750–1820, was also associated with a population explosion and had very different social effects from the sheep farming and deer forests that dominated the next century and were associated with clearance and the maintenance of land as empty space. This contrast bred a bitter antipathy of crofter and small farmer towards large farmer and landowner without in the least rocking the secure foundations of landed power. Crofter, farmer, forester, laird, who farm by land and sea, plant trees and shoot, survive today as the social, economic and political backbone of Highland life, perhaps more reconciled towards one another in the second half of the twentieth century than for two hundred years, and this by the pressure of mutual dependence on various forms of agricultural subsidy and mutual antipathy towards non-traditional forms of land use.

The third attitude is to regard land as a resource for industry, which I also classify as traditional as it has fascinated entrepreneurs and planners, mainly outsiders, since Sir George Hay's ironworks in Wester Ross in the seventeenth century. By and large, it has been a sad disappointment; the eighteenth century was littered with failed ironworks, mines and textile undertakings, and the twentieth century with pulp mills, aluminium smelters and rig yards. Even hydropower and the nuclear industry never realised the immense hopes vested in them by Tom Johnston and the post-war Secretaries of State, as agencies for reviving the Highlands by providing power for light industry. The main importance of those hopes for the Hydro Board historically lie in explaining why Scotland, alone of advanced industrial countries, had no land designated as national parks until 2002.

All three of these traditional attitudes to land usage (and the first two, related to agriculture and private sport, are much the most important) leave out of account any interest an outsider might have in using the Highlands, or any scenic or scientific value the land might have. They are fundamentally informed by an ancient way of seeing nature as resilient and there to be exploited and the land as providing a way of making a living or pursuing a private pleasure.

The next three attitudes I have classed as post-romantic, as none of them would have been thinkable before the age of Ossian. Number four in my list is an attitude that regards the land as an invigorating obstacle course: if mountains, then as something to climb up, ski down, ramble along, or if water, taken to swim in or sail upon – a location for mass, popular and public sports, originating in the outdoor sporting movements of around the 1880s, growing with the rambling and youth movements of the first half of the twentieth century, but only really taking off in the age of mass leisure and popular access to the car from the 1960s to the present day. Sporting recreational use by outside visitors is clearly one of the main pressures on the Highlands now, bringing new calls for free access and countervailing protests from, for example, the Scottish Landowners Federation.

The fifth attitude is a century older and follows Thomas Gray and Wordsworth in seeing unspoiled landscape as refreshment to the spirit, so something to be maintained in its entirety and contemplated in its tranquillity. There was much emphasis on this throughout the nineteenth century, but only on the part of intellectual romantics with no influence on public policy; in the first half of the twentieth century they allied very effectively with those who held the fourth view (of land as an arena for public recreation) to put pressure on government to declare national parks. If they failed then, they succeeded in the creation of the National Trust for Scotland in 1931 with a membership that grew from only 1000 as late as 1946 to 21,000 in 1961 and 180,000 by 1989;[97] and succeeded further in political lobbying for creation of a Countryside Commission for Scotland, which came about in 1968 with power to designate National Scenic Areas with a substantial degree of protection for some of the most beautiful parts of Scotland.

The sixth and last attitude is to regard land not only as a recreational and scenic resource for man, but also as a refuge for plants, birds and animals seen as interesting and worth preserving for their own sake. This attitude had, as we have seen, very little public emphasis in Britain, and particularly little in the Scottish Highlands, until after the Second World War. Then it grew dramatically. This can be demonstrated by the transformation of the RSPB from a rather inward-looking and elitist organisation into a great popular body. In 1954, when it still had fewer than 7,000 members in Britain, the decision was taken to open a regional office in Scotland. In 1958, under RSPB care, a pair of ospreys returned to Speyside after an absence of half a century, and the following year over 14,000 people visited the hide

[97] Figures kindly provided by the National Trust for Scotland.

EXPLORING ENVIRONMENTAL HISTORY

overlooking their nest; ten years later, 37,500 visited the same site during the summer season. By 1987, the UK membership of the RSPB was 440,000, of whom 24,000 (or 5.4 per cent of the total) were Scottish.[98] This nevertheless compared with 160,000 members of the NTS, indicating the extent to which Scottish support for scenic and historic conservation was still much deeper than support for nature conservation alone.

The three attitudes towards the Highlands that I have called post-romantic were all held predominantly by outsiders to the Highlands, and thus became the main reason for the tourist trade servicing sport, scenery admirers and nature lovers, which flourished exceedingly in the age of mass leisure and car ownership, 1960–90.[99] 'Gentlemen the Tourist' was the title of a biting attack by Neil Gunn in 1937,[100] and ever since there have been those who have expressed doubts about tourism, often on the grounds that it is, in some way, an inappropriate means of earning a living compared to fishing or agriculture. The fact that so many incomers (the 'white settlers' of the popular contemporary Highland phrase) cater to tourism while so many whose families have lived immemorially in the Highlands restrict themselves to traditional land use forms increases the tension and the ambivalence. The fact remains that in areas like Arran and Speyside tourism has become the main economic mainstay of life in the last thirty years, and is likely to become more important everywhere as subsidised farming declines.

Highland society, in fact, has come to accord a differential scale of respect to the three post-romantic attitudes. Mass sport and recreation may be unacceptable to some landowners and farmers because of what they regard as trespass, but it is seen to provide jobs and does not interfere with traditional land use in other ways. Scenic appreciation is less welcome if it gives rise to criticism of forestry policies and fish farming, or limits development in National Scenic Areas. Nature appreciation seems to have been the least welcome of all, at least since the 1981 Wildlife and Countryside Act gave teeth to the Nature Conservancy Council to interfere substantially with the

[98] Samstag, *For Love of Birds*, figures on p. 149. Scottish Tourist Board, *Tourism in Scotland* (Edinburgh, 1969). Other details kindly provided by the Royal Society for the Protection of Birds. In May 2008, RSPB membership in Scotland was 77,933, or 7.4 per cent of UK membership. In February 2008, NTS membership was 302,000.

[99] But of course the expansion was well under way before 1960. In 1951, 215,000 overseas visitors came to Scotland, rising to 470,000 by 1959. By 1960, 5 million visitors a year, British and foreign, were using Scottish tourist facilities: *Natural Resources in Scotland*, pp. 702, 714.

[100] *Scots Magazine*, 26 (1936–7), pp. 410–15. See also the *Newsletter of the National Trust for Scotland*, 2 (April 1949), pp. 7–8, where the spokesman expressed himself 'not convinced that the future of the Highlands lies in a vastly increased tourist trade. I doubt the Highlanders desire or have the ability to be the manager of a big hotel, a first-rate restaurateur, golf-course attendant or AA Scout'.

rights of landowners, forestry companies and farmers to do what they liked with their property by declaring SSSIs. It is of course quite likely that a single person will combine many attitudes in one; a visitor who climbs a mountain, admires the view and identifies with pleasure a peregrine falcon on the crags has all three of the post-romantic attitudes. It is also quite likely that some will conflict, for example sport and nature conservation in the Cairngorms since the 1950s, most recently and most dramatically in the disputes over the Northern Corries and the funicular railway. In this case, traditional land use attitudes gather round the sports interests to present a common 'insider' Highland front. 'People in the Highlands and Islands do not want or need to be told what developments can, or cannot, take place on their land by those often many miles removed from the physical reality', said Sir Robert Cowan, chairman of the Highlands and Islands Development Board in a Cambridge University Union debate in April, 1990.[101] 'Any developments which threaten such a landscape are of international interest: we can hardly ask the Third World to stop felling their rainforests if we cannot look after our own heritage . . .' responded the leader writer of *Country Life*, sympathetic to the Save the Cairngorms Campaign.[102] These quotations (and their origins) neatly encapsulate the insider versus outsider conflict, the traditional versus the post-romantic. And that conflict has become, in the last twenty years, a severe, endemic and destructive element in modern Highland history.

So where does this leave us? This lecture has, I hope, demonstrated how all our attitudes towards the land and nature have a history, indeed, how they have come to be twisted and directed by history. But a last and more important point might be made. The situation at the moment is a potential disaster because too many on the development side cannot see either that the Highlands belong to a wider British society than seems to be visible from Inverness, or that man is an animal along with other animals on this fragile planet. On the other side, too many see economic change only in simple and emotional terms of man's encroachment on nature's kingdom. One might well consider that the only possible way forward is to bring to our aid a holistic ecology concerned with developing overall land-use strategies according to the strains the land can bear, treating man as an animal who fits in with the natural world instead of trying to smash through it. There are certainly no magic formulae or simple solutions, but a return to the spirit of Patrick Geddes in the 1890s or of Frank Fraser Darling in the 1940s would be to recover a Scottish tradition that was eclipsed and lost in its day, but which is needed now as never before in our history.

---

[101] As reported in a leading article, *Country Life*, 24 May 1990.
[102] Ibid.

POSTSCRIPT

Following publication of the lecture, I received two particular pieces of constructive criticism. Jim Hunter took me to task for too little consideration of pre-existing green consciousness among the native Gaels, and for not understanding its continuity and current power. In a remarkable book, he assembled a powerful body of literary evidence to demonstrate a deep and longstanding indigenous feeling for nature.[103] He shows how the Irish hermits of the eighth to tenth centuries produced nature poetry marked by a spare and beautiful directness of observation unmatched in the European tradition, arguing that to the end of the Middle Ages Ireland and the Scottish Highlands formed a single culture region where similar poems were still composed. He also shows how the Highland Gaelic poets of the eighteenth century, like Alasdair MacDonald and Duncan Ban MacIntyre, though influenced by the poetry of James Thomson, wrote poetry surprisingly free of the 'sentimental and didactic bias' that marked the work of Thomson, Walter Scott and the English romantics. Closest to the spirit of their tenth-century forebears, and to John Clare and George Crabbe among English contemporaries, they were amazingly direct and sensitive in their observation of the natural world. If the Gaelic poetry of the nineteenth century does not quite rise to the literary heights of its predecessors, it is equally independent in rejecting the fashionable admiration of the wild and desolate, and contrariwise regrets the desolation of once inhabited places cleared by sheep and emigration.

Possibly Hunter overstates the absence of didactic bias: Scottish Gaelic praise poetry in the bardic tradition often uses the 'pathetic fallacy', where a feature in nature echoes a human situation – nature laments in sympathy with human emotions, a curlew wails for the death of a hero, and so forth. And the importance of Lowland influence on Highland poetry, especially from the eighteenth century, cannot be denied. Nevertheless, he establishes an unassailable case for the existence of a distinctive Highland poetic description of love of nature, and one appreciated not so much by the elite as by the common people.

Jim Hunter then asks himself why, if ordinary Highlanders had an innate sensitivity to the natural world (reasonably assuming that their appreciation of the poets indicated an appreciation of what they were describing), they so often appeared hostile to twentieth-century nature conservation. To this, he replies that nature conservationists too often behaved in an insensitive,

---

[103] J. Hunter, *On the Other side of Sorrow: Nature and People in the Scottish Highlands* (Edinburgh, 1995).

top-down way that riled the crofters and the farmers. Maybe they did, but my declared purpose was to examine the context of a movement to conserve the Highlands from depredation upon its scenery and ecology and, for whatever reason, Gaelic green consciousness had not, up to 1990, played much of an obvious part in bringing active conservation about.

It might perhaps be argued, as Highland farmers and crofters themselves sometimes argued, that they were naturally more gentle in the treatment of the environment than others because of their innate appreciation of the environment. But on average they were no slower than others in similar positions to adopt the latest agricultural technologies and practices where they could, whatever the cost to the natural world. The Highlands hardly provided the terrain for the full gamut of methods that barley barons and factory farmers applied in Suffolk. And if there is an innate Celtic love of nature, it certainly has not borne fruit in Ireland, the most backward of all western European nations in respect to conservation.

Perhaps the Highlander who did most in the public realm to further the cause of conservation was the remarkable naturalist Seton Gordon, born in Deeside in 1886, to whom a significant place in my lecture should certainly have been allotted. The influence of his books – with his straightforward multi-faceted appreciation of the Highland countryside, and of his inspiration for a generation of conservationists and ecologists in the third quarter of the twentieth century – was considerable, as Hunter rightly emphasises.

The second criticism came from Camilla Toulmin of the International Institute for Environment and Development (and an expert on Africa) in a private letter, part of which is quoted below:

In your final section, perhaps one could add a seventh attitude which is emerging, which regards land and other environmental assets as providing fundamental services, without which our systems of production would be unable to survive and we, ultimately, would find our existence impossible. I am thinking of work being carried out by environmental economics, which attempts to value such services. For example, wetland areas provide valuable functions in absorbing water and regulating stream flows, influencing the volume and reliability of water supplies to many downstream users. Tree cover similarly enables greater absorption of rainwater and, hence, slows run-off and reduced the risk of erosion. Consequently, benefits from increasing tree cover should include reduced siltation of rivers and dams. Only now are we starting to realise the importance of the services provided by the natural environment. Such a realisation has probably only taken place because of instances where damage to the environment has been so great that such services are no longer being provided. Somehow, it is always easier to identify a system which has become unsustainable, than one which is sustainable. Identification and explicit valuation of the services provided by the natural environment could enable sounder policies to be designed, and more comprehensive decisions to be made which took account of the broader costs and benefits from particular activities.

These observations were correct and far-sighted; one of the conservation advances in the fifteen years since that letter was written has been a much greater appreciation of the Scottish environment in precisely these terms.

Another has been a recognition in Scotland of an aspect of conservation politics to which Camilla Toulmin also referred in her letter – 'the need to rethink the relationship between land users and outside experts or professionals . . . to enable [the latter] to listen properly to what local farmers have to say, and try to build on issues, problems and priorities identified by local people . . .' In this, she said, the African debate was considerably in advance of the Scottish one.

Toulmin and Hunter, coming from entirely different perspectives, had therefore identified a weakness in the Scottish situation in the 1980s, which was actually rooted in the history of green consciousness: because as a programme for action rather than as a background sentiment it was rooted in science and outsider middle-class activism, it was indeed top-down rather than bottom-up.

When Scottish Natural Heritage came into being in 1992, it was charged with having 'regard to the desirability of securing that anything done, whether by SNH or any other person, in relation to the natural heritage of Scotland, is undertaken in a manner which is sustainable.' This was the first time the concept of 'sustainability' had been used in a British statute. However, when the Board of SNH attempted to comment on the sustainability of a new Forth road crossing, it was slapped down by the Scottish Office and told to stick to the narrower remit, conservation of the natural heritage, or risk a cut in funding. The radical wording was by no means equalled by radical government intentions, and an opportunity was lost.

Magnus Magnusson, chairman of SNH, and Roger Crofts, chief executive, also set about trying to improve dialogue with local communities, thus to reduce tension and increase effectiveness. A limited number of academic studies suggest they had some success,[104] but the arrival of the European

---

[104] A. J. Whitehouse, 'Negotiating small differences: conservation organisations and farming in Islay', unpublished University of St Andrews Ph.D. thesis, 2004; E. B. Johnston, 'People,

Habitats Directive after 1995 threw them back onto an obligatory site-based and top-down approach. Now that these European sites have been designated, the organisation has resumed the search for consensus, and for treating the wider countryside as the main focus for action.

Community-based initiatives and bottom-up input into agendas have, therefore, become the favoured approach since the lecture was first delivered. They are not easy to achieve and there is a certain risk in achieving them: a community might prefer, for local reasons, to embark on a course of action damaging to a national or international interest (for example, oppose the reintroduction of a species valued nationally but locally judged to be potentially harmful to economic interests). In practice, however, this risk is not great, since national and international rules and regulations (top-down in origin) now limit what a local community can do. It is not fashionable to say so, but it is probably just as well that the history of green consciousness worked in such a way that top-down preceded bottom-up. Top-down set the rules of the game; bottom-up may be the best way to implement them. New agendas, which will be essential in a changing world, can probably only effectively be set by a good working relationship between bottom-up and top-down. Fraser Darling, with his zeal for ecological science and his vision of man's central place in nature, would surely have agreed with this.

Interestingly, in May 2008, the Scottish government, advised by Scottish Natural Heritage, decided to go ahead with the contentious reintroduction of beavers following public consultation in which 70 per cent of the participants favoured the return, although 57 per cent of those living close to the reintroduction site did not.[105] So even an SNP administration, publicly committed to listening to local people, is prepared to act top-down, providing the science supports the decision and that it has the backing of a wider public.[106]

---

peatlands and protected areas: case studies of conservation in Northern Scotland', unpublished University of Aberdeen Ph.D. thesis, 2001.

[105] *The Herald*, 26 May 2008.

[106] For further studies and reflections on some of the themes of this lecture, see T. C. Smout (ed.), *Nature, Landscape and People since the Second World War* (Edinburgh, 2001), being the proceedings of a conference in Edinburgh to celebrate the fiftieth anniversary of the National Parks and Access to the Countryside Act of 1949.

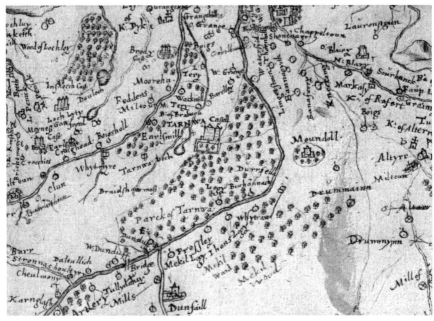

Fig. 3 Timothy Pont's map of Moray, from just before 1600, shows a well-wooded landscape round Darnaway, with an enclosed 'parck' and the 'Mekil Wood' that for centuries had provided high-quality oak timber for royal ships and palaces. Between 1760 and 1810, these woods were transformed and modernised by planting twelve million trees, mainly Scots pines. Reproduced by permission of the Trustees of the National Library of Scotland.

CHAPTER 3

Exploiting Scottish
Semi-Natural Woods, 1600–1850*

The central question in this chapter is how far the exploitation of Scottish woodland, 1600–1850, was, in modern terms, 'sustainable'. Some preliminary explanations are in order. Sustainability is essentially a twentieth-century term, the meaning of which has evolved gradually, beginning with the American Progressive notion of a 'maximum sustainable yield' applied to fisheries and forestry in the era of Theodore Roosevelt and Gifford Pinchot, but more recently expressed by the Brundtland Report as a universal ideal of sustainable development which 'meets the needs of the present without compromising the ability of future generations to meet their own needs'. This has subsequently been glossed as necessarily involving firstly, the maintenance of biodiversity within ecosystems and secondly, the acceptance of such development by local societies.[1]

The general ideal enunciated by Brundtland would have been immediately recognisable by early modern society. It is no more than a general notion of wise traditional use at peasant level 'adhered to because traditionally they were the only guarantee of survival'[2] or of 'good stewardship' at proprietorial level. It resonates, for example, with the laws of entail in England and Scotland, which limited the rights of landed heirs to sell property acquired by their forefathers. The Earl of Lauderdale, political economist and Scottish landowner, knew all about sustainability when he wrote in 1804 that 'the common sense of mankind would revolt at a proposal for augmenting wealth by creating a scarcity of any good generally useful and necessary for man'.[3]

---

* This is a substantially revised version of a paper written with Fiona Watson, 'Exploiting Semi-natural Woods, 1600–1800', in T. C. Smout (ed.), *Scottish Woodland History* (Scottish Cultural Press, Edinburgh, 1997), pp. 86–100. Thanks to Fiona Watson for her very considerable input into the original, and for permission to reuse from Avril Gray of Scottish Cultural Press, Unit 6, Newbattle Abbey Business Park, Newbattle Road, Dalkeith, www.scottishbooks.com.

[1] United Nations Conference on Environment and Development, *Agenda 21* (Rio de Janeiro, 1992).

[2] M. Redclift, *Sustainable Development: Exploring the Contradictions* (London, 1987), p. 150.

[3] James Maitland, eighth Earl of Lauderdale, *An Inquiry into the Nature and Origin of Public Wealth and into the Means and Causes of its Increase*, 2nd edn (Edinburgh, 1819), p. 44.

On the other hand, the subsequent gloss would have been incomprehensible in an earlier age. The notion of preserving biodiversity was meaningless in a society that took for granted that man had been given dominion over every growing plant and living creature, to use or to extirpate them at pleasure. Similarly, while the notion that change had to be socially acceptable might have worked in societies where peasants had effective control over land management, it scarcely determined the behaviour of the Scottish landowners, whose absolutist rights of property could legally override the customs of their dependants to a degree exceptional in Europe.[4] We might therefore say that if early modern woodland management practices met the needs of the time without compromising those of future generations, that was probably an intended consequence, but if they happened to preserve biodiversity or to be socially acceptable to the commonality, that was an accidental by-product. Furthermore, to be realistic, it is perfectly possible and even likely that a management regime which maintained and increased the 'maximum sustainable yield' of timber products would, to some degree, both damage biodiversity and offend the local population. This was certainly true in late eighteenth-century Scotland.

What woods are we talking about? It has been estimated from Roy's military survey of circa 1750 that woodland at that time covered about 4 per cent of the Scottish land surface, running from 2–3 per cent in the Lowlands to 7 per cent in parts of the Highlands.[5] The accuracy of these calculations is open to question because of the short-comings of Roy's map, which omitted about half the woods that appear from their shape and character to have been ancient semi-natural woods in the First Edition of the Ordnance Survey (1856–1891) – the Ordnance Survey also itself omitted a fair number of woods present on Roy and still there today. Roy also underestimated the size of some of the woods, a fact which can be checked by reference to contemporary estate maps. A revised estimate, also based on reconsideration of Sir John Sinclair's compilation of statistics for 1814, puts the area of Scottish land surface under semi-natural wood at the end of the eighteenth century at around 7 per cent, with perhaps another 2 per cent for plantations.[6] This is probably an upper-bound estimate, and will include a great deal of land that had little more than scrub growing on it, or was open to

---

[4] T. C. Smout, 'Landowners in Scotland, Ireland and Denmark in the Age of Improvement', *Scandinavian Journal of History*, 12 (1989), pp. 79–97.

[5] A. C. O'Dell, 'A view of Scotland in the middle of the eighteenth century', *Scottish Geographical Magazine*, 69 (1953); J. M. Lindsay, 'The commercial use of woodland and coppice management', in M. L. Parry and T. R. Slater (eds), *The Making of the Scottish Countryside* (London, 1980), p. 272.

[6] T. C. Smout, A. R. MacDonald and F. Watson, *A History of the Native Woodlands of Scotland, 1500–1920* (Edinburgh, 2005), pp. 64–7.

wood pasture – under some definitions not woodland at all. Even this more generous figure would still rank Scotland among the less-wooded countries in Europe.

All of this wood would have been 'semi-natural', composed of native species growing by natural regeneration on their original sites, but modified by the pressure of human beings and their domestic stock over millennia – in many cases since the Neolithic. Timber scarcity appears in the Scottish Lowlands from the fourteenth and fifteenth centuries and thereafter became a talking point for outside travellers, from Aeneas Sylvius around 1430 to Dr Johnson in 1773. There was clearly a striking regional imbalance between the Lowlands and the Highlands, with much more woodland in the latter region.

The reasons for this imbalance were not simply that the Lowlands were more fertile, had a denser population, and more towns, so that ground that was bearing trees came to be required for cultivation. Southern England and France provide examples of much more extensive woods in more populated countries, and much of the Scottish Lowlands in fact consisted of bleak poor quality ground capable of carrying a timber crop but unsuitable for cultivation. One better alternative explanation is the abundance in Scotland of alternative fuel and building material: peat was very widely available, and coal was noted as a common domestic fuel from the fifteenth century, initially dug in shallow bellpits or from outcrops that needed no more capital than a pick, a horse and a cart. Stone was everywhere present, an impediment to cultivation that could as well be put in the walls of a house as in a clearance dyke, and turf could also be used to build houses and outbuildings. There were certainly many things for which wood was indispensable, for example building crucks and rafters and most agricultural tools, but a community could make do with relatively less than in most parts of Europe.

Towns, with their large elite houses and churches, sometimes even with palaces and cathedrals, were always big consumers of building wood – Edinburgh lavishly so from the sixteenth to the eighteenth century. But Scottish burghs were all coastal, and satisfied their needs not from Scottish sources but from Norway and the Baltic, where timber was cheaper, better quality and much more accessible. It was a couple of days sail with a good wind to the 'Skottarhandel' area of south-west Norway, where deals could be loaded in the deep fjords 'at the woods', as contemporaries put it. Compare that to the prospect of beating up to the Moray Firth and waiting at Speymouth for wood rafted down from Abernethy or Rothiemurchus, of a quality usually considered to be fit just for the roughest building work where it was out of sight. There was no contest.

A further explanation of the scarcity of woods is the relative importance, especially in the Border counties, of extensive sheep farming. It has been calculated that the Scottish wool clip in the late fourteenth century was the produce of over 2 million sheep.[7] The same forces that kept the English Downs free of scrub kept the Southern Uplands of Scotland green and heathery. The effect of the rising profitability of sheep ranching was reflected in the manifest decline of woodland in the royal and baronial hunting 'forests' of the Borders, such as Ettrick Forest in the fifteenth and sixteenth centuries, and has been traced on monastic lands in Dumfries and Galloway between the Urr and the Nith, where grants of pannage before around 1180 (implying herds of pigs in oak forest) gave way abruptly to grants of pasture for sheep.[8]

Nevertheless, it would be quite wrong to imagine the south of Scotland as either completely devoid of woods or lacking a tradition of woodland management. The woods were simply more restricted in scale than in most parts of Europe. The main species in their composition were oak, with birch, willow, hazel and alder as important in most places, and elm, ash and sycamore as locally significant. The last-named was an exotic from southern Europe, but planted in Scotland from the late Middle Ages both around houses and in woods: there is sycamore coppice in woodland at Cults in Fife that may be 400 years old.

A native tradition of Scottish woodcraft analogous to that described for southern England and France[9] emerges from a study of woodland contracts in the seventeenth and in the first half of the eighteenth centuries.[10] The specialist terminology used suggests their antiquity: 'haggs', for divisions of a wood to be cut at any one time, derives from an Old English and Norse word for the hewing down of trees, shared with the north of England; 'rice' was from another Old English and Norse word for twigs and brushwood used as wattle on the fencing to exclude animals from the coppice; 'grain' was from a word for branch; 'hain' meant to hedge or protect, and so on.

The main provisions in Lowland contracts embody three elements.[11] Firstly, the trees were to be cut at the base with the expectation that they

---

[7] A. Grant, *Independence and Nationhood: Scotland 1306–1469* (London, 1984), p. 79.

[8] M. L. Anderson, *A History of Scottish Forestry* (London, 1967), vol. 1, pp. 164–81; R. Oram, *The Lordship of Galloway* (Edinburgh, 2000), pp. 251–8, and *pers. comm.*

[9] O. Rackham, *Ancient Woodland: its History, Vegetation and Uses in England*, 2nd edn (Dalbeattie, 2003); R. Bechman, *Trees and Man: the Forest in the Middle Ages* (New York, 1990).

[10] National Archives of Scotland: Register of Deeds. For detailed references to these deeds cited in the following text, please see Smout, MacDonald and Watson, *Native Woodlands*, *passim*, or e-mail the author at tcs1@st-andrews.ac.uk.

[11] See Smout, MacDonald and Watson, *Native Woodlands*, ch. 7. For a local study, T. C. Smout, 'Managing the woodlands of East Lothian, 1585–1765', *Transactions of the East Lothian Antiquarian and Field Naturalists' Society*, 26 (2006), 41–53.

would regrow – that is, they were to be coppiced (pollarding and shredding may well have been practised but are not mentioned in the deeds). The early rules of coppicing in Scotland included stipulations to cut the stool 'raised and smoothed as to prevent the diminishing thereof by water' (or alternatively not to be cut 'dished', or to be cut 'with ane ascent and descent' – with an upward and downward stroke); other contracts stated that it had to be cut four inches from the ground, and that the roots must not be damaged in any way. The cutting season was stipulated: for example [he] 'binds himself not to cut any timber in the flee month [or worm – caterpillar] which is called 15 days before Lammas and other 15 days after Lammas yearly' (between 16 July and 15 August). Often the season for oak began in April and ended in July to facilitate stripping the tanbark, worth 40 per cent or more of the total value of the tree. Other trees than oak might be called 'winter wood'. Small trees or withies were to be left, sometimes defined as those that could not be drilled by a small tool – could not bear a 'wormlick bore'. Very frequently it was a regime of coppice-and-standards, where a certain number of 'maiden' trees, growing straight from seed, were to be left: for example, 'John and Alexander bind themselves . . . to leave in the several haggs of said woods as they cut the same yearly, 100 reserve or maiden trees proportionally, the most part whereof to be 8 inches round at 6 quarters high from the root'.

Secondly, a system of staggered cutting, and sometimes of cutting on a clear rotation, came to be established. Small woods might be felled in one season, or divided into a small number of haggs to be cut in turn, but the larger ones such as Branxholm and Mortoun near Langholm were on a ten-year fell. It does not of course follow that the woodcutters here would begin again after the ten years: the woods might well then rest another decade. The Wood of Kincardine on the Montrose estate, however, was sold in 1704 on a twenty-four-year rotation basis, the 'wood to be divided in as many equal portions as there are years allowed to John and he to cut but one of these portions in a year', and in 1722 the same estate sold the 'third part of the woods of Monteith now ready to be cut . . . in seven years by equal divisions', suggesting a twenty-one-year rotation. Arrangements were made for cleaning the haggs of brush and rubbish after cutting, for regulating the grazing of horses in the wood while work was continuing, for setting up woodcutters' houses, and so on. In the Torwood in 1740, the cutters were to 'thicken the haggs so cut with young trees from the nurseries in the yards of Torwoodhead'; but this was exceptional: regeneration was the usual way of replenishing the felled divisions.

Thirdly, careful arrangements were frequently made for establishing and maintaining enclosures round the wood, and round the individual rotation haggs within the wood in order to exclude damage from animals. This was the most important provision of all. Thus, at Innerkip in 1704 the woodmen felling a hagg had to 'build ane sufficient fencible chardyke of stake and ryce so as to keep out all horses, sheep or other cattle and that by 14 July yearly . . . and to keep up the same'; at the Wood of Kincardine in the same year they had to 'uphold the peallings [palings] within the wood during all the time of haining'; at Kippen in 1712 they had to 'give as much timber as will serve for pealling of the outside and indivisions of the said woods'; at Mortoun in 1722 they had to 'allow so much walling and top hedging as shall be found necessary and convenient for Sir Patrick for the feall [turf] dykes for inclosing the said woods and making cross dykes betwixt the haggs or faulds of the same'. There is a very strong impression, even in the Lowlands, that use by animals and woodland management were regarded as complementary. A wood was valuable shelter, particularly in winter, and it yielded some grazing in the open grassy patches. On the other hand, animals would damage the regrowth ('spring') and had to be excluded from the haggs for a variable number of years to allow it to grow tall enough to be out of reach.

When management fell below standard, there was awareness of what to do. When, in 1715, it was discovered that the woods on the Buccleuch estates in Ettrick forest were in poor shape, the chamberlain was instructed to make a particular inspection, to appoint proper foresters, to prosecute those who cut, stole or barked wood, and since 'we are informed there is a good appearance of young springe in severall places . . . specially near Newark where wood grows naturally and every spring makes a pretty good show, but destroyed for want of incloseing', the chamberlain was to mark out the places most proper to be enclosed with fences, 'sufficient and sensible against beasts'. Three years later, detailed recommendations on the Ettrick woods called for the birch and alder to be fenced and coppiced; the oak and ash to be thinned if kept for standards, or coppiced with their stocks being carefully cut at the ground and the springing thinned; and the hazel to be thinned, enclosed and coppiced or weeded out to make room for more valuable wood.[12]

Good practice – that is, stipulations about enclosure and felling – increased between 1600 and 1750, not least in the seventeenth century. In the first half of the seventeenth century, about 65 per cent of Lowland woodland deeds contained such stipulations, rising to about 85 per cent in 1650–1700, and by 1750–9 this had increased to about 95 per cent.

As in the English and French systems, woods managed under these regimes would perhaps have been indefinitely sustainable. Doubts have been thrown on the long-term viability of coppicing however, particularly under

---

[12] A. H. H. Smith, 'A history of two Border woodlands', in T. C. Smout (ed.), *Scottish Woodland History* (Edinburgh, 1997), pp. 152–3.

conditions like those in the west of Scotland where continuous systematic removal of woody material, combined with the effects of heavy rainfalls on acid soils, might have a deleterious effect. Good woodland management could also give way under other pressures on land use. The monastic house of Coupar Angus in the fifteenth and sixteenth century managed woodland at Campsie by dividing it into quarters, maintaining haining regulations and appointing local tenants to have the responsibility of being foresters. However, the demand for grazing land was such that animals were allowed into the enclosures, and in 1551, part of the wood was eventually reported as 'waistet and destroyt';[13] two centuries later, there was no trace of it. It is not only good management, but continuity that ensures sustainability. Very seldom have we encountered instances of a wood being grubbed out to turn the area into cultivated ground, though there was an instance in 1743 near Tulliallan in Clackmannanshire.

The Highlands had retained appreciably more wood than the Lowlands, with a different balance of trees. Birch was the most widespread species in most regions, dominant particularly in Wester Ross and Sutherland. Oak dominated in Argyll and Perthshire, but there were also often oakwoods, especially on south-facing slopes, as far north as Inverness and Ross-shire. Scots pine, which did not occur naturally south of the Highland line, was dominant, generally in mixed woods with birch, on poorer soils in two different habitats. The eastern pinewoods comprised the relatively large forests of the Spey, the Dee and of the Beauly catchment, along with a few straths to the north. The western pinewoods ranged from near Ullapool to Glen Orchy in Argyll, but were more scattered and normally on north-facing slopes. The main difference was that in the east the pinewoods regenerated from seed relatively easily in a drier climate and on more sparsely vegetated ground, whereas those in the west depended more on catastrophe, such as fire or wind-throw, to create space in the deeply mossy or peaty ground sufficient for seed to germinate. Other common trees in the Highlands were hazel (usually coppiced), ash (where there was limey ground), alder and willow (where it was wet), gean, rowan, holly and aspen. The last three were probably much commoner than today, as they tend to be selectively browsed by domestic stock and deer. It is not unusual to find early reference to woods of 'birk and hollyn'.

There are many problems in trying to ascertain the extent of woodland in the Highlands before the middle of the eighteenth century. The first maps are those of Timothy Pont from the close of the sixteenth century; on the whole, they do not suggest that much more wood was present then than there was a

[13] C. Rogers (ed.), *Rental Book of the Cistercian Abbey of Coupar Angus* (London, 1879–80).

century and a half later, but they are (for the most part) field sketches rather than finished productions.[14] Robert Gordon of Straloch also left unpublished rough manuscript maps; these have tree symbols over much wider areas than Pont, but there is some uncertainty about his first-hand knowledge of the ground. When Blaeu in Amsterdam published worked-up versions of the Pont maps and of others emanating from Gordon in 1654 they often had more woodland than on most of the Pont sketches, but less than the Gordon maps; this could have been as much for decorative as for cartographic reasons.[15] Much more detailed manuscript maps of the upper Dee (around 1703) and of parts of the Great Glen, Easter Ross and the Beauly catchment area (1725–30), suggest a distribution of woodland recognisably similar to that of later times, though not without some loss – on the Mar Lodge estate, for example, woods in Glen Derry and Glen Lui are now much reduced.[16]

Some of the ambiguity may derive from uncertainty as to what to count as a wood, especially in the Highlands. Many of the early symbols seem to indicate scattered groupings rather than continuous wood, strikingly so in Pont's map of Glencoe and Mamore, which is also annotated 'many fyrre woods here alongs' (probably the tree symbols were added by Gordon). This was an area that became almost totally denuded of Caledonian pine after 1750. Then there are several descriptions from the late eighteenth and early nineteenth centuries of hardwood scrub that appeared above the level of the heather in summer, only to be eaten back by domestic animals in winter when they came down from shielings on the hills. Williams spoke of 'a great many thousand acres' of such land forming 'a rich stool of oak in a deep soil' between Speyside and the Atlantic sea lochs; Monteath of 'many thousand acres of land that was formerly carrying Natural Woods [that] have of late years been left unenclosed and set aside for pasture land', particularly in Argyll.[17] Perhaps these were the vestiges of much larger natural woods that had flourished in the Middle Ages but were now in their final stages of decay.

Examination of eighteenth-century woodland contracts often reveals differences in management practices between the Highlands and the Lowlands. Before 1750, there is generally a striking lack in the north of specific details to protect the wood by regulating coppicing, determining rotation and, above all,

---

[14] J. C. Stone, *The Pont Manuscript Maps of Scotland: Sixteenth-century Origins of a Blaeu Atlas* (Tring, 1989); I. Cunningham (ed.), *The Nation Survey'd: Timothy Pont's Maps of Scotland* (East Linton, 2001).

[15] Smout, MacDonald and Watson, *Native Woodlands*, pp. 48–56.

[16] Royal Commission on the Ancient and Historical Monuments of Scotland, *Mar Lodge Estate, Grampian: an Archaeological Survey* (Edinburgh, 1995).

[17] J. Williams, 'Plans for a Royal forest of oak in the Highlands of Scotland', *Archaeologica Scotica*, 1 (1974), p. 29; R Monteath, *Miscellaneous Reports on Woods and Plantations* (Dundee, 1827), pp. 54–5.

an absence of stipulations about fencing. Contracts may specify that certain trees (for example, the 'firs' or pines) should not be cut or that trees below a certain size are not felled. Often the restrictions were in dangerously vague terms: for example, 'Arthur and Roger bind themselves to cut down the trees according to the common and reasonable custom and at the proper times and seasons'. In this last case (relating to a contract between Irish adventurers and the Earl of Breadalbane) the landlord later declared that the damage done by the contractors had been intolerable to the value of the estate and the interests of tenants, but he had no redress. The line between the detailed Lowland contracts and the looser Highland ones sometimes follows the geographical line quite exactly: in Perthshire a contract relating to Innerpeffery on the Lowland side specifies coppice restrictions, haggs and fencing, but one relating to Logierait and another to Faskally on the Highland side specifies coppice regulation more vaguely, and haggs and fencing not at all. It is the absence of enclosure after cutting that is particularly striking, though in some areas such as Lochtayside oakwoods were 'emparked' apparently from an early date.

The explanation is not that Highland woods were considered of no value. In cultural and aesthetic terms, woods ranked high even in the sixteenth and seventeenth centuries when (at least according to Keith Thomas) they were regarded poorly in England. The contrast is between epithets like 'delectable', 'fair', 'fair and tall', 'beautiful to look on', 'pleasing' in Scotland, and 'dreadful', 'gloomy', 'wild', 'desert', 'uncouth', 'melancholy' in England.[18] Woods were also regarded as extremely useful to the Highlanders in economic terms. Even stone buildings needed wooden crucks and cabers, and in the most forested areas the walls of entire houses were also made from wattle or planking. Prestige buildings like castles were likely to use local, not imported, wood before 1750. All sorts of equipment, from ploughs and harrows to mill machinery, needed wood: temporary fencing demanded wattle frames of hazel, fish traps and creels were similarly constructed, holly boughs were cut for fodder, fir candles (the most resinous parts of pine trees) were used for light, bark was used for tanning leather and preserving fishing nets. The practical uses of wood in Highland society were infinite.[19]

There are better explanations for the neglect of enclosure. Firstly, woodland was relatively much more plentiful in most places in the Highlands (the islands excepted) than it was in the Lowlands, so a need to husband the resource was not so clearly seen, and the physical difficulty of enclosure was

---

[18] K. Thomas, *Man and the Natural World* (London, 1984); ch. 9 below.
[19] Smout, MacDonald and Watson, *Native Woodlands*, ch. 4; I. F. Grant, *Highland Folk Ways* (Edinburgh, 1995); H. Cheape, 'Woodlands on the Clanranald Estate: a case study', in T. C. Smout (ed.), *Scotland since Prehistory: Natural Change and Human Impact* (Aberdeen, 1993), pp. 50–63.

much greater. Second, pine does not spring again from the bole when cut, and birch does so only weakly, so the full elaboration of Lowland practice was less applicable outwith the main oak-growing areas. Third, there was an important use of woodland as shelter for animals, the bottom end of a transhumance system that took them to the hill pastures, the shielings, in summertime. In places, animals remained all year in the wood. Grazing that got out of control holds the key to much of their ultimate decline: the failure, in fact, of sustainable use.

Although all woods, including those subject to coppice in the Lowlands, were used to pasture stock at some time in their felling cycle, especially in the Highlands there were stretches of upland wood pasture (such as Glen Finglas) where the system of exploiting the ground was fundamentally different. The trees were often spaced well apart, to facilitate the growth of nutritious grasses between them; they were not coppiced but pollarded (the contemporary phrase was 'cut high') just out of reach of the animals – except the goats, unfortunately, which stood on their hind legs or even climbed into the trees. Pollarding does not seem to have been done in Scotland to provide leaf-fodder, as in Scandinavia or the English Lake District, but to harvest wood for fuel in places where the peat bogs were distant, and also for construction purposes. The distinction between a wood and a wood pasture was not absolute: at low altitudes the wood might be dense, but thin to a scatter of trees as it went up the hill. Nor did it mean that the trees were incapable of regeneration from seed, as herd boys could keep the cattle away from sensitive areas. But it was a fragile system, capable of collapse if there was undue pressure from numbers of animals or severity of weather. The commonest trees to be pollarded in these upland wood pastures are probably alder (on wet slopes), hazel and birch, and frequently rowans and other small trees grow in their crucks and forks where leaves and moss accumulate; this provides the bizarre spectacle of trees growing upon trees.

Even where the trees were not widely spaced and not obviously pollarded, but ordinary upland birch woods, they were also often used as wood pasture. The light canopies of birch, and the ability of the tree to improve the quality of the ground beneath it by leaf-fall, facilitated the growth of grass within the wood in a way that could not happen in a lowland oakwood. Grazing animals and Highland woods were inseparable.

It is important here to distinguish between the proximate and the ultimate cause of Highland deforestation. Some contemporaries, and most historians, were quick to blame a series of obvious culprits when they saw the forest felled. For the pinewoods, chief among them were English speculators, beginning on the Spey at Abernethy with the lease of the woods to Captain John Mason in 1631. It is not known whether Mason actually cut any timber, but the York Buildings Company purchased 60,000 trees in the forest from the owner in 1726 and set about its exploitation with great vigour, floating the wood down to the sea in rafts for the first time, building an iron furnace that operated briefly, and erecting sawmills. The company came to grief through over-extravagance, but sales and felling continued. Higher up the Spey at Glenmore and at Rothiemurchus there was intensive felling late in the eighteenth and in the early nineteenth centuries, partly for ships' timbers and partly for deal planking; contemporaries after the close of the Napoleonic Wars spoke of 'devastation' in these ancient forests. But the fact remains that each of them are still there, possibly diminished in size (it is difficult to be sure) but undeniably sustained. A similar point could be made about the Black Wood of Rannoch, and the forests in the Beauly catchment area, where we know of serious attempts at exploitation, but also from pre-1760 maps that they cover much the same ground now that they did 250 years ago. It could be cogently argued that the very value put on the timber gave the lairds the incentive to maintain them. Some of the baron court penalties for misuse of the wood were draconian on Speyside: in 1693 burning heather too close to the woods was punishable by having the culprit's ear nailed to the gallows (it happened to children more than once), and in 1722 re-offending by stealing wood three times was allegedly punishable by death.[20]

It is also true, however, that in these eastern pinewoods regeneration was very much easier, under both normal circumstances and either after fire or after disturbance of the ground, than it was in the west and north. In the west, it is easy to find comparable acts of woodland exploitation, but many of the woods here are certainly now much smaller than formerly, and some have been reduced to small remnants or virtually disappeared. Natural woods here were perfectly capable of regenerating vigorously, especially after fire, but could not do so if the seed trees had been removed – especially if the eighteenth-century form of removal, by horse, was insufficiently destructive of the thick western moss and heavy vegetation to allow fresh-shed seed to come up through it. What was sustainable in the east might not be so in the west.

And of course it was always true that if heavy levels of browsing were allowed, whether by deer, sheep, goats or cattle, a pinewood, like any other wood, could not regenerate. In the east, fires in forests like Glentanar were usually followed by regeneration, but 7,000 sheep came to be pastured in it and fires were not then followed by new growth.[21]

---

[20] Smout, MacDonald and Watson, *Native Woodlands*, chs 8, 11, 12; ch. 4 below; H. M. Steven and A. Carlisle, *The Native Pinewoods of Scotland* (Edinburgh, 1959).
[21] Steven and Carlisle, *Native Pinewoods*, p. 96.

The history of oak woodland in Argyll and elsewhere in the south and west Highlands is comparable.[22] Here the traditional villains are usually named as the ironmasters, beginning with Sir George Hay in the early seventeenth century in Wester Ross, and continuing with Irish and, especially, English interests in the eighteenth century in Invergarry, Loch Etive and Loch Fyne. Many accounts stress the depredations that the charcoal burners are supposed to have wreaked for this handful of furnaces, up and down the western sea lochs, though Lindsay demonstrated the inherent implausibility of such widespread damage, as well as showing that the longest-surviving of the furnaces, on Loch Fyne, was associated with an increase in local woodland, not a decline.[23] The exploitation of the oakwoods for tanbark, however, was more prolonged, more widespread and heavier than the cutting for industrial fuel. It was tanbark that brought Irish adventurers to the woods – for example, of Lochiel – even before the Act of Union and which continued to attract Scottish exploitation on a large scale well into the nineteenth century. The explanation for the early Irish interest lies in British legislation that forbade the export of Irish cattle to mainland Britain; this drove the Irish to slaughter their animals at home, barrel the beef for ships provisions for the Transatlantic trade (English and French ships used Cork as their main provisioning port) and tan the hides themselves. By the end of the seventeenth century, Irish supplies of oak bark were insufficient.

In the early days – before the second half of the eighteenth century – the cutting of the Highland oakwoods appears from contracts to have been largely unregulated, but as the price of charcoal and bark increased all this changed. By the time of the agricultural reports of the 1790s, it was normal in the main oak-growing areas accessible to transport, to have coppice-management on twenty- to twenty-five-year rotation, followed by fencing to protect the regrowth for at least five years, and sometimes permanently. Further north – and more distant from the market – there were still extensive patches of oak 'all shamefully abandoned, after every cutting, in the same neglected condition', and even in Argyll critics in the early nineteenth century still spoke of thousands of acres of oak regrowth abandoned to grazing.[24] Although the biodiversity of the woods must have been considerably modified as birch, hazel, willow and holly were discouraged in the interest

---

[22] Ch. 5 below.

[23] J. M. Lindsay, 'The use of woodland in Argyllshire and Perthshire between 1650 and 1850', unpublished Ph.D. thesis, University of Edinburgh, 1974; J. M. Lindsay, 'Charcoal smelting and its fuel supply: the example of Lorn furnace, Argyllshire, 1753–1876', *Journal of Historical Geography*, 1 (1975), pp. 283–98.

[24] J. Robertson, *General View of the Agriculture of the County of Inverness* (London, 1794); Monteath, *Miscellaneous Reports*, pp. 53–4.

of the more valuable oak, in most respects and despite reservations this was the hey-day of sustainable broadleaf forestry in the Highlands.

All this was brought about by a management revolution. Until the middle of the eighteenth century, day-to-day exploitation of the Highland woods was loosely co-ordinated by the local baron courts, which fined the tenants for such offences as cutting greenwood without permission, for 'cutting high', for burning birch within a wood or practising muirburn too close to a wood. Occasionally they suggested an interest in other good forest practice, such as systematically planting trees. The seventeenth-century baron court books of Glen Orchy in 1621 provide a good illustration: every year, each tenant or tacksman of a merkland was to set out six young ash or sycamore trees, and every cottar three, and to transplant them again to the 'maist commodious pairtis of thair saidis occupatioun' when they were grown on. The estate gardener provided them at two pennies each, which seems unfair as the trees on the land presumably still belonged to the laird.[25] Local tenants, as in the Lowlands, were generally appointed to be foresters, but in the Highlands they were not obviously concerned to regulate grazing (except occasionally that of goats), since a prime use of woodland was for winter shelter for stock.[26]

When in the eighteenth century this system came under the critical eye of the agricultural Improver and land steward, they were extremely critical of its effects; they found the regulations neglected, the woods pillaged for fuel, bark, and construction timber, and animals everywhere eating the regeneration. Their answer was to replace local control by a professional factor or specialised forester from the estate head office and then basically to introduce the Lowland system of wood management, with a particular emphasis on enclosure and on ending the former freedoms for the tenants to take what they needed from the wood. The reforms produced complaint of real hardship among tenants who found themselves deprived of critical pasture and shelter for their stock at the most difficult time of the year. They also became short of timber for construction and everyday use: thus in Argyll where wood had been so freely available that farm houses were 'in some respects more commodious than in many other parts of the Highlands', a few years later the reservation of the woods for charcoal and tanbark production had created in districts such as Kintyre 'a great discouragement to the farmer', who then needed to buy his timber from Norway.[27] One serious consequence of the managerial change was that the link in much of

[25] C. Innes (ed.), *The Black Book of Taymoth* (Edinburgh, 1855), p. 35.
[26] See also F. Watson, 'Rights and responsibilities: wood-management as seen through baron court records', in Smout (ed.), *Scottish Woodland History*, pp. 101–14.
[27] J. Robson, *General View of the Agriculture of the County of Argyll* (London, 1794), p. 58; J. Smith, *General View of the Agriculture of the County of Argyll* (Edinburgh, 1798), p. 132.

the Scottish Highlands between woodcraft and farming was broken forever. When the fences round these oakwoods again fell into decay, as the price for tanbark and charcoal collapsed in the later nineteenth century, they were again used for pasture with very damaging results, and generally with little attention to any other value they might have to the farmer.

How far were the Improvers' criticisms of the traditional looser controls justified, and, if so, why had an apparently unsustainable system of peasant husbandry been allowed to continue for so long? Consider, first how the traditional grazing system would have worked. Those modern enemies of natural regeneration in Scotland, deer and sheep, were a much smaller problem than now. Deer numbers were a fraction of what they are today, and in some areas where regeneration no longer occurs solely due to their browsing, their absence was a matter of concern to the landlord anxious for better sport.[28] Sheep numbers similarly were low, and the size of the animal small; until the great invasion of the Highlands by Cheviot and Blackface sheep, mainly in the first half of the nineteenth century, the Highland sheep was diminutive and kept by the tenants mainly for their own subsistence.

The principal browsing animals of the traditional Highlands were horses, goats and cattle. Horses (though much more numerous than today) were probably not a problem as they are light feeders on trees; goats certainly were a problem, and, though also kept mainly for subsistence, were extremely numerous – 100,000 goat skins were sent from the Highlands to London in one year at the end of the seventeenth century.[29] The Improvers were extremely critical of the damage they did, and most estates had effectively extirpated them from their tenants' flocks by the end of the eighteenth century. The greatest volume and weight of beasts, however, must have been the traditional Highland 'black cattle', which were kept both for the peasants' own use and to meet their rent obligations to the landlord. Cattle grazing had a plus and a minus side. On the plus side, the cow has heavy feet that punch holes in a mossy sward and thus favour regenerating trees, especially in the wetter west; it also feeds on grass with a tearing motion, which also disturbs the ground, and has runny dung that penetrates at once into the ground as fertiliser. Sheep and deer, by contrast, are too light to make much impact, nibble the sward closely like mowing machines, and have hard dung, much of which dries and evaporates on the open hill. On

---

[28] J. S. Smith, 'Changing deer numbers in the Scottish Highlands since 1780', in Smout (ed.), *Scotland since Prehistory*, pp. 89–97; A. Watson, 'Eighteenth century deer numbers and pine regeneration near Braemar, Scotland', *Biological Conservation*, 25 (1983), 289–305.

[29] *Scottish Studies* had four articles dealing with 'Goats in the old Highland economy' viz. B. R. S. Megan, 7 (1963), pp. 201–8 and 8 (1964), pp. 213–17; M. Campbell, 9 (1965), pp. 182–6; T. C. Smout, 9 (1965), pp. 186–9.

the minus side, however, if cattle are too numerous they can create as much damage to the wood as any other animal simply by eating everything that comes up. In the eighteenth century, the complaints were of heavy and increasing grazing, and the observations numerous and damning.

One such was a very interesting comment in Sutherland in 1812, where the reporter told of a 'remarkable alteration on the face of this part of the country in the course of the last twenty years', the widespread decay of the natural broadleaf woods with which the straths were once covered. The reporter was torn between blaming climatic change and animal grazing. He said that 'naturalists aver' that severe frost and snow in April and May had caused the destruction, but as the main species was birch, which endures much colder climates than Sutherland, the explanation of a run of cold springs seems at first unlikely; though such is the oceanic character of the Scottish environment compared to that of Scandinavia, that if an increase in precipitation were combined with an increase in grazing pressure in a cold spring, the effects could be serious for regeneration. Certainly the grazing was heavy. Until very recently every farmer had kept a flock of 20 to 80 goats, and 'the constant browsing of black cattle' made it 'not surprising that the [natural] oak is nearly gone'. At any rate,

> it is a well known fact, that in the straths where these woods have already decayed, the ground does not yield a quarter of the grass it did when the wood covered and sheltered it. Of course the inhabitants cannot rear the usual number of cattle, as they must now house them early in winter, and feed, or rather keep them just alive, on straw; whereas in former times their cattle remained in the woods all winter, in good condition, and were ready for the market early in summer. This accounts for the number of cattle which die from starvation on these straths, whenever the spring continues more severe than usual; and this is one argument in favour of sheep farming in this country.

In a corroborative letter, the minister of Kildonan adds the details that formerly the animals were outwintered until January but must now be taken indoors in November, and that the replacement of the 'fine strong grass with which the woods abounded' by coarse heather had led to a 'degeneracy of black cattle in the parts that were formerly covered with wood'.[30]

It would be hard to find a better contemporary description of the knock-on effects on an ecosystem of unsustainable woodland management,

---

[30] J. Henderson, *General View of the Agriculture of the County of Sutherland* (London, 1812), pp. 83–6, 106, 176.

but it was not the first hint in this direction. For example, as early as 1753, a tenant on the Mackenzie of Seaforth estates in Wester Ross had written:

> I hear from sensible honest men that other places in the country besides my tack have now less wood and more fern and heath than formerly, so that cattle want shelter in time of storm (as we never house any) and their pasture is growing more course and scarce. I know of severall burns that in time of a sudden thaw or heavy rain are so very rapid that they carry down from the mountains heaps of stone and rubbish which by overflowing their banks they leave upon the ground next them for a great way and by this means my tack and other are damaged and some others more now than formerly.[31]

Erosion and increased stream-flow are just what would be expected from deforestation, in addition to deterioration of the herbage. We know of no external exploitation here to explain the losses. Interestingly, it was in these northern counties of Sutherland and Ross where most complaint came in the nineteenth century of deteriorating pastures, then attributed to sheep grazing.[32] It may have longer and different roots.

The eighteenth century was marked by two novel features in the history of the Highlands – an unprecedented increase in human population, and an unprecedented increase in the numbers of black cattle as the external trade to the Lowlands and to England grew. Human population of the five Highland counties grew by well over one half between 1755 and 1841, and cattle exports grew by at least fourfold between 1720 and 1814; not all would be Highland animals, but unquestionably many were. Then in the first half of the nineteenth century, cattle were largely replaced by even larger numbers of sheep. This pressure blew apart the balance between farming and woods. It could be argued that earlier practices of grazing wood pasture at light levels had been at least semi-sustainable. Unfortunately, the eighteenth-century change was both subtle and insidious, and land management traditions were too inflexible to reduce the pressure of animals. This kind of thing must be common in the Developing World today. The impact of all this on biodiversity extended beyond the trees themselves – the wolf was exterminated probably before 1700, and the largest woodland bird, the capercaillie, by 1780. But above all, the traditional ways of treating the Highland woods began to become much more damaging because there simply were too many people with too many animals.

[31] National Library of Scotland MSS. 1359, 100.
[32] J. Hunter, 'Sheep and deer: Highland sheep farming, 1850–1900', *Northern Scotland*, 1 (1973), pp. 199–222.

An instructive example comes from Deeside in the 1740s, in a lawsuit where the rights of a tenant of Farquharson of Invercauld to cultivate on the edges of a forest were challenged by the feudal superior, Lord Braco, who argued that it would prevent the regeneration of his pines which needed space outside the existing wood to seed successfully, as they would not do so within its bounds unless it was cut or blown over. The growth of farms in an area hitherto only lightly used by man and animal was depriving the woods of their natural means to gradually change their boundaries within a matrix of open moor, the mechanism by which they had traditionally survived.[33] Pressure of people meant that sustainability was threatened.

From this trap there were three ways out. The first was for local communities to resist the pressures and to continue to accord the traditional natural woods enough respect and value for them to survive. That this could at least occasionally happen is demonstrated in Assynt, west Sutherland, where the woods that were in existence around 1770 are virtually all in existence today, particularly those in the neighbourhood of settlements.[34] A very similar situation pertains at Sleat on Skye.[35] The second was for the eastern pine forests to come under commercial pressure in the nineteenth century, and for their owners to respond by better management, including enclosure of a wide area against sheep, as happened at Rothiemurchus on Speyside. The third was for similar pressures to inspire the enclosure and care of the oak-woods of the southern Highlands, many of which became exceptionally well tended in the period 1780–1850. However, the last two remedies proved of short-term value when the price of Scottish wood products collapsed in the nineteenth century. Most even of the eastern pinewoods were failing to regenerate until recently due to pressure from deer numbers, now being brought under some control in forests such as Abernethy. Both oak and birch woods are generally still under heavy pressure from deer and sheep. That many have survived over the last century and a half is due not so much to enlightened management as to the ability of trees, as organisms, to endure for very long periods, providing they are not physically removed from the landscape. Even so, many semi-natural woods have been destroyed: we have lost 40 per cent of Highland birchwoods since the Second World War. But just allowing them to survive, but not to regenerate, which was usually the extent of the most generous twentieth-century management over much of Scotland, is of course not a sustainable woodland policy in the long run.

---

[33] J. G. Michie, *Records of Invercauld* (New Spalding Club, Aberdeen, 2001), pp. 124–53.

[34] R. Noble, 'Changes in native woodland in Assynt, Sutherland, since 1774', in Smout (ed.), *Scottish Woodland History*, pp. 126–34.

[35] Alan MacDonald, *pers. comm.*

Fig. 4 Scots pines growing at the edge of Glenmore Forest, 1935. Glenmore was one of the last of the ancient pinewoods to be intensively exploited, falling to the axes of Dodsworth and Osborne of Hull between 1784 and 1806. Twenty-five years later, it was reported to be regenerating vigorously. It takes more than clearfell to wipe out a forest in the Cairngorms. Photographer: Valentine Collection. Courtesy of St Andrews University Library: JV-A2604.

CHAPTER 4

The Pinewoods and Human Use, 1600–1900*

The native Scottish pinewoods have, since medieval times, been subject to several kinds of human use. They have in many cases been deer forests, reserved for elite hunters, and they have all been wood pastures, used by the farmers' cattle, sheep, horses and goats. They have been a timber and fuel resource for local people, and the subject of exploitation by the external market. Only since the twentieth century have they been widely admired and visited by outsiders for their beauty, biodiversity and historic significance, though the roots of this admiration lie with the Victorians.[1] Each one of these uses has left its mark on the woods, along with the under-lying effects of climate. Each one has also varied in character and impact with the changing centuries.

The use of the deer forest, for example, changed over time. Initially animals were driven by a circle of men ('tinchel') towards a funnelled enclosure ('elrig') where they could be killed by dogs and with weapons, but since the eighteenth and particularly since the nineteenth century, the sport involved stalking on the open hill by one sportsman and his ghillie, and the woods were used only for sheltering the animals. We know nothing of the ecological impacts of the first phase, but the second was and remains an important source of the failure of the woods to regenerate naturally. As Louis Boppe, of the Nancy Forest School, said on a visit to Scotland in 1881:

> As foresters of the Continental school, accustomed to live among forests regularly managed, and having for their object the

---

* This is reproduced from a paper given at the Caledonian Pinewood symposium at Drumnadrochit and published, amended, in *Forestry*, 79 (2006), pp. 341–9. Thanks to the Institute of Chartered Foresters for permission to reprint with minimal alterations.

[1] R. A. Lambert, 'In search of wilderness, nature and sport: the visitor to Rothiemurchus, 1780–2000', in T. C. Smout and R. A. Lambert (eds), *Rothiemurchus: Nature and People on a Highland Estate 1500–2000* (Edinburgh, 1999), pp. 32–59.

production of timber, we had no little difficulty in understanding the widely differing motives which activate forest cultivation in this country. Everywhere we found the forests fenced on all sides with walls and hedges ... We learnt that these costly enclosures were erected, not for keeping out the cattle and deer, as in the Jura, but for keeping them in! It appeared to us like shutting the wolf in the sheepfold.[2]

The use of the pinewoods for wood pasture, providing shelter and grazing for domestic stock was extremely widespread in the period before 1800, and had probably been so from time immemorial. Thus the Forest of Mar around 1760 was referred to as 'the best out-pasture for all kinds of cattle to be found anywhere in Scotland', and at Rannoch in 1780 the tenants complained that if the Black Wood was fenced for timber production they would lose their best pasture and only wintering ground.[3] After that time, proprietors interested in exploiting the pinewoods primarily for timber sales began systematically to exclude stock, but there were many others who had no interest in wood sales or found little opportunity to pursue them and continued to allow woods to be used for grazing. As time passed, however, sheep increasingly replaced cattle, and contemporaries observed the consequences. As David Nairne put it in 1892:

In the beginning of the nineteenth century the institution of sheep rearing on a large scale had a distinct effect upon the Highland forests. The area under wood ceased its natural expansion, the young seedlings being all eaten up, while the herbage got so rough that there was not a suitable bed for the seed to fall in. On the other hand, black cattle, which formerly occupied the hills and valleys in large numbers, were favourable to the production of forests, as they kept the herbage down and trampled the seed into the ground, the result being that wherever they fed in the proximity of a wood a luxuriant crop of trees invariably made its appearance.[4]

[2] Quoted in M. L. Anderson, *A History of Scottish Forestry* (London, 1967), vol. 2, p. 337.
[3] J. G. Michie (ed.), *Records of Invercauld* (New Spalding Club, Aberdeen, 1901), pp. 143, 145, 153; J. M. Lindsay, 'The use of woodland in Argyllshire and Perthshire between 1650 and 1850', unpublished University of Edinburgh Ph.D. thesis, 1974, p. 137.
[4] D. Nairne, 'Notes on Highland woods, ancient and modern', *Transactions of the Gaelic Society of Inverness*, 17 (1892), p. 220.

In some respects no doubt the absence of cattle made the heart grow fonder, for there were many eighteenth-century observations of the damage that grazing animals of all kinds did to timber production in woods. In 1798, John Smith noted how the remnants of old pinewoods in high glens shed seed driven over the moor by the storms, from which 'a beautiful plantation rises up in spring' only for 'this precious crop, the hope for future forests', to be destroyed when the cattle are driven up in spring.[5]

As long as grazing and hunting was a dominant use of the woods, we may expect it to have had a profound impact on their structure. In place of the close, straight-grown trees desired by the forester, one would expect a wood of clumps and granny pines, open-grown and park-like in character, as many traditional Scots pinewoods are. There are a number of contemporary indications that seem to confirm this. Around 1590, Timothy Pont gave a sketch of the woods running along the Gruinard River to Loch na Sealga, writing on the map that it was a 'mechtie parck of nature'. He described in Glencoe 'many fyrre woods heir alongs' but he (or more likely Robert Gordon of Straloch) used quite scattered tree symbols on part of this map as though trying to convey a certain openness.[6] Estate maps at the Doune of Rothiemurchus dated 1762 and 1789 also used a mapping convention that suggests that even that forest was not, on much of the ground, as densely wooded as Roy's military survey (for instance) shows.[7] When the Hull merchants bought adjacent Glenmore in 1783, they 'observed many coarse crooked trees standing in the forest which we understand were of little or no value, but as we intend to build ships they will be useful to us'.[8] Most explicit of all is an account of 1827 by J. E. Bowman, approaching the Spey from the north, over the Slochd:

As we dropped down into Strathspey, the road kept winding among large detached and scattered groups of trees, principally pines, which had an appearance so novel and striking, and so different from the artificial plantations in England, that we felt convinced they were the Natural Pine Forests we had read of. Trees of all sizes grew intermixed, though at a sufficient distance for each to assume its full character

[5] J. Smith, *General View of the Agriculture of the County of Argyll* (Edinburgh, 1798), p. 146.
[6] J. C. Stone, *The Pont Manuscript Maps of Scotland: Sixteenth Century Origins of a Blaeu Atlas* (Tring, 1989), pp. 38, 91.
[7] Doune MSS, Rothiemurchus.
[8] J. Skelton, *Speybuilt: the Story of a Forgotten Industry* (Garmouth, 1995), p. 24.

and receive the full influence of the sun and air, and for the larger
ones to throw out their broad declining arms in graceful sweeps to a
great extent.[9]

Trees of this character had value for local use, as curved timber was preferred for house crucks and for ploughs, harrows, the cas-chrom and other
tools; while strongly curved or forked timber was favoured to tie the crucks
together at the top.[10]

However, open-grown trees in sparse woods were certainly not the only
form of forest structure. Animals were readily controlled even without
fencing in the centuries before compulsory education took the child herders
off to school (from around 1870), so it would not be difficult to reserve
portions of a wood to grow tall straight trees for specific local use. There
is, for example, archaeological evidence of this in Castle Grant, spanned
in the sixteenth century with very fine pine timber up to eight metres long
and thirty-five centimetres in diameter, and also evidence in the account of
the vanished seventeenth-century towerhouse of Macdonell of Glengarry
which apparently also used timbers eight metres in length. These must have
been close grown.[11]

A relatively small number of pinewoods had a reputation for being
able to supply masts, which also surely implies tall, close-grown trees.
Conaglen in Ardgour was one, supplying the Lowlands and Ireland, but
the best known in the seventeenth century was the wood at Strathcarron,
where an English traveller in 1677 saw pines 'very high and straight, but
of no great substance, about a man's fathom'.[12] The Royal Navy of the
Cromwellian and Restoration period was interested in buying these as an
alternative to Riga masts, but in the event found the sample supplied too
small and unsuitable, 'sorry stuff' in the words of authorities at Chatham
dockyard.[13] Strathcarron nevertheless for a time renewed a trade in naval
masts that earned a bounty after 1707, and when the owner of Rhidorroch,

---

[9] J. E. Bowman, *The Highlands and Islands: a Nineteenth-Century Tour* (Gloucester, 1986), p. 161.

[10] I. F. Grant, *Highland Folk Ways* (London, 1961), pp. 101–6, 142–9; G. D. Hay, 'The cruck-building at Corrimony, Inverness-shire', *Scottish Studies*, 17 (1973), pp. 127–33.

[11] U. Lee, 'Native timber construction; Strathspey's unique history', *Scottish Woodland History Discussion Group Notes*, 7 (2002), pp. 23–9; H. Cheape, '"A few summer sheilings or cots": material culture and buildings of the '45', *Clan Donald Magazine*, 13 (1995), p. 68.

[12] P. H. Brown, *Tours in Scotland 1677 and 1681 by Thomas Kirk and Ralph Thoresby* (Edinburgh, 1892), p. 36.

[13] T. C. Smout, A. R. MacDonald and F. Watson, *A History of the Native Woodlands of Scotland* (Edinburgh, 2005), pp. 323–31.

near Ullapool, asked why he was not getting the same price for his wood as his friend in Strathcarron, he was told that by contrast his was inferior, not being 'narrow', since 'they have been made havok of at pleasure, not only by all your own people in this country but also by the neighbourhood without much control'.[14]

Another wood, subject to strict control, and which became known for tall straight trees, was Ballochbuie, where the owners built sawmills to process the wood and floated much to market at Aberdeen. Thomas Pennant saw trees there in 1769, 200 years old and up to 'twelve feet in circumference and near sixty feet high, forming a most beautiful column' and also said that nearby this 'ancient forest' was 'another, consisting of smaller trees, almost as high, but very slender'.[15]

Taking wood for local use mostly involved small timber, for construction, tool manufacture, fuel and light for the farmer and the cottar. Big straight timber would be of no use to them. As the reconstructed houses at the Highland Folk Museum at Newtonmore show, homes were made of various materials, including stone foundations, turf walling and heather thatch, but lavish use was made of small-size wood, and where pine was available it was obviously used, though never exclusively – birch and alder were often easier to obtain and just as useful. Because there was no advantage for local people in allowing trees to grow to great size, needs were satisfied by either cutting limbs off granny pines or harvesting trees of small dimensions, with or without the permission of the laird. On the estate of Cameron of Lochiel in the eighteenth century, the wood at Loch Arkaig was preserved for the laird's own use, but that of Glen Loy was for the use of his tenants.[16] The strange, cut-about appearance of the oldest Scots pines there seems to testify to such exploitation. In Rothiemurchus, for example, activity of this kind was called 'cutting the green wood' and punished by the baron court.[17] This was not untypical, but it made sense to come to an arrangement whereby tenants could help themselves without interfering too much in the estates other interests, since those who needed timber like this also paid the rent.

The use of woodland for fuel varied from place to place. Dead wood was always a useful bonus for the locals, and estate regulations allowed it

---

[14] M. Clough, 'Early fishery and forestry development on the Cromartie estate of Coigach, 1660–1746', in J. Baldwin (ed.), *Peoples and Settlement in North-west Ross* (Edinburgh, 1994), pp. 235–7.

[15] T. Pennant, *A Tour in Scotland, 1769* (Edinburgh, 2000), pp. 82–3.

[16] Smout *et al.*, *Native Woodlands*, p. 139.

[17] Doune MSS, Rothiemurchus, 1705–6, 1772–3.

to be taken even when there was strict prohibition on taking green wood. Many parishes had peat supplies that could replace wood fuel, and if it was convenient and of good quality, the time and effort needed to cut the peat was often less than that taken to harvest living wood. In nineteenth-century Strathspey, however, even before the railways, the minister of Alvie reported that 'peat was so distant from the parish and produced at so enormous a price that coal from Inverness was actually cheaper'.[18] Wood fuel was then essential, but birch was preferred over pine by both the laird (who might regard it as a weed of the forest) and the tenant (who preferred less resinous wood in his hearth).

On the other hand, for light, 'fir candles', the torches and splinters made to illuminate ordinary homes, had to be cut from Scots pine on account of the high resin content. The inhabitants of the Black Wood of Rannoch and those of Glen Tanar were two groups who made a small living from cutting torch material from living trees and selling it at a considerable distance at the markets of Perthshire and Angus. In Abernethy in 1763 complaints about the abuse of the wood included 'a most pernicious custom of cutting pieces from the body of large and thriving trees for candles and other uses, which wood for candles has been sold and sent out of Strathspey'.[19] The mention of 'other uses' than candle fir can only be a reference to tenants helping themselves to boughs and crucks from living pine.

What determined the structure and ecological value of the woods? Structure may be thought of as naturally a function of soils, drainage and altitude, but it was also likely to be affected in our centuries by the market. If the wood was far from a usable river or sea loch, or if the market for timber externally was weak, the wood's prime value was for animal shelter and grazing. Then we might expect the development or accentuation of an open structure with many spreading trees. If it was in a favoured location close to a shipping or floating point, we might expect more protection and encouragement of naturally close-growing trees. Where, as at eighteenth-century Rothiemurchus, there was a lively market for small pole timber for relatively local use, this may have kept the average size of the trees to within a dimension that could be comfortably cut by local axes and carried by horseback. Where there was an opportunity to use or sell for sizeable construction or shipbuilding timber, as at Strathcarron, there might be an attempt to encourage trees to grow much larger.

---

[18] D. Nethersole-Thompson and A. Watson, *The Cairngorms* (Perth, 1981), p. 22.

[19] G. A. Dixon, 'Forestry in Strathspey in the 1760s', *Scottish Forestry*, 29 (1976), pp. 40–1.

The market for animals and human demographic trends were also relevant factors. Under circumstances where there was pressure from cattle, or cultivation on the moor outside the wood, there might be interruption of natural regeneration. How pines regenerated away from their own shade was a process well understood in the eighteenth century, as illustrated during a lawsuit over the Forest of Mar in 1758–60:

> It is known to your Lordships, that, as fir-woods do not spring from the root, but are propagated by the blowing of the seed in the grounds immediately adjacent to the old woods, or in the openings, where they have freedom of air, these highland fir-woods are not fixed to a particular spot, but gradually shift their stances.[20]

Other consequences of human activity were also independent of the market. The gathering of deadwood would certainly have left the forests with less fallen and decaying material than modern ecologists would consider optimal for the benefits of biodiversity. On the other hand, the cutting into living wood for lighting and building material left – from the laird's point of view – an infuriating number of damaged pines, but insects, fungi and birds might have found that a gain.

From around the second quarter of the eighteenth century a livelier market for wood resulted in a quickened interest in the exploitation of the pinewoods by outsiders. This had been foreshadowed in the seventeenth century by a certain number of events in the eastern forests, like the interest shown by Captain Mason in the Grant woods of Speyside around 1630, and the slightly more sustained interest shown by the Royal Navy in the masts of Strathcarron in 1652 and in 1665–9. In the west, there was more sign of activity in a trade in deals and masts from a number of scattered localities including Loch Maree in the north, Ardgour, Loch Leven and Loch Etive. It is doubtful if any of this seventeenth-century activity had much impact, except perhaps to extract some of the biggest trees, especially in the west. At first, eighteenth-century activity was marked by the high rate of business failure, generally before much sustained felling had taken place: this appears to be true in, for example, the Loch Arkaig woods, those of Strathglass, and in Rothiemurchus. At Abernethy, where the woods still appeared to outsiders to be virtually untouched when the York Buildings Company arrived a full century after Captain Mason, the

---

[20] Michie (ed.), *Invercauld*, pp. 142–3.

subsequent felling operations had not gone very far before the company declared bankruptcy.[21]

Altogether, the depredations of outside exploitation on the woods before around 1790 must be considered very limited, with one or two exceptions such as the catastrophe of the felling of the woods of Glen Orchy, of which more in a moment. Another was the fact that the selective removal of the largest and straightest timber, and therefore of many of the oldest trees in close-grown groves, was probably widespread at all the accessible sites. If this is combined with a general increase in pressure on the woods through the rise of rural population and their stock (with cattle prices buoyant) throughout the Highlands in the course of the eighteenth century, it is probably true to say that late eighteenth-century woods would have had a younger average age-structure than those of 300 years before. They were also probably more affected by heavy grazing by cattle, horses, sheep and goats both inside the woods and on the moors around, inhibiting regeneration.

What happened in the next half of the century was much more dramatic than anything before. From 1790 onwards, the lairds largely took over the exploitation of their own forests from outside speculators. The one important exception to this was the lease of Glenmore by the Hull firm of Dodsworth and Osborne in 1784 that felled the wood over the next 20 years for shipbuilding timber. Otherwise, the main exploiters were the Grants of Grant at Abernethy, the Grants of Rothiemurchus at Rothiemurchus, the Dukes of Fife and Farquarson of Invercauld in upper Deeside, the Earl of Aboyne at Glen Tanar, and so forth. In each case, faced with rising timber prices during and for a time after the wars with France, the owners cut heavily into the woods until there was little more saleable timber left to harvest.[22] To the casual observer the results were catastrophic. As P. J. Selby saw it in 1842:

> The indigenous forests of Scotland, which formerly occupied so large an extent of its territory have, within the last sixty years, been greatly reduced, in consequence of the demand for pine timber occasioned by the difficulty of obtaining wood from the Baltic during the late wars: some, indeed, are nearly obliterated.[23]

[21] Smout *et al.*, *Native Woodlands*, esp. ch. 8.

[22] Ibid.

[23] P. J. Selby, *A History of British Forest Trees, Indigenous and Introduced* (London, 1842), cited in I. Ross, 'A historical appraisal of the silviculture, management and economics of the Deeside forests', in J. R. Aldhous (ed.), *Our Pinewood Heritage* (Farnham, 1995), p. 142.

It was not quite a modern clearfell, because often the structure of the woods in terms of tree spacing and varied age would not lend itself to that, but it was much more than the intermittent selective felling of previous centuries. And it was accompanied by a quite novel attitude to alternative uses of the wood. In the past, hunting, grazing and local timber use had been broadly compatible: lairds did not want to hunt all the time, and providing the numbers of stock were not too great, they were happy to share the deer pasture with the farmers' animals, and in recognition that farmers needed wood supplies to be able to continue to farm and to pay rent, they were also content to allow them to take it, subject to certain safeguards. Now, however, the profit from the timber in the wood overruled all other considerations. Deer were, in the biggest forests, often excluded as far as possible. Farmers, likewise, were ordered to remove their stock, and sometimes told to buy what wood they needed from the estate rather than to help themselves. Sawmills became larger and organised more on an industrial scale, as at Rothiemurchus. Forestry became much more professional. A wood was there to produce timber – everything else was a distraction.

The effects of this half century were not as disastrous as Selby implied. No forest was 'nearly obliterated', though the desolation that followed the quasi-clearfell struck many with horror. The most perceptive, however, also noticed how rapidly regeneration was recurring, and within a few decades the woods seemed much as before.[24] It remains unclear exactly what ecological damage this episode did, for example to the distribution of the characteristic flora, some of which may be highly susceptible to episodes of clearfell, or to insect life in the forests. The highly fragmented distribution of many species may owe a lot to the devastation of this period, particularly as it was mimicked again by the near clearfells that took place in many forests in the two World Wars.

By 1850, however, that episode was behind, and for the remainder of the nineteenth century there was little money to be made from pine timber in the old woods, unless, perhaps, as in Strathspey, a railway was being constructed nearby and a market for wooden sleepers suddenly appeared. Lairds reacted in two ways. For a few, the answer was to concentrate on the best-known modern forestry techniques, to plant pine close, to try foreign strains, to weed out birch where you could and to take a pride in your woods, since only by going for quality and good husbandry could you expect to break even: the Lovat estates were the leaders in this approach,

---

[24] T. Dick Lauder, in W. Gilpin, *Remarks on Forest Scenery and Other Woodland* (Edinburgh, 1834), vol. 1, p. 175.

but their woods were mainly plantations.[25] For most, the answer was to relinquish the effort to grow profitable timber, to allow the woods to recover naturally since they might indeed be valuable again one day, but to allow them to go over primarily to a sporting use or to be part of a sheep run. To be a deer forest was the fate of such woods as Rothiemurchus and Ballochbuie, while at Abernethy the end of active forestry in 1878 resulted in the clearance of at least 104 people to make a deer reserve that employed five. To be a sheep run was the fate of Rhidorroch and several other woods. Glentanar was said to have 7,000 sheep running in the pinewoods.[26] This was the period when Louis Boppe arrived in Scotland to express astonishment at the policy of fencing animals inside the woods, and when David Nairne believed that the switch from cattle to sheep had led to a failure of natural regeneration. The nineteenth century had been an age of a new intensity, successively in both felling and grazing, with biodiversity effects that are unknown in detail and may have been considerable, but which certainly did not amount to obliteration.

As far as the records allow us to say, most pinewoods that existed at the start of the seventeenth century exist at the present time, and many of these are of comparable dimensions to what they apparently were in the seventeenth and eighteenth centuries. But a number of pinewoods have indeed been obliterated since 1600, or reduced to small remnants, and it is instructive to see where and when this took place.

It is impossible to give a fully comprehensive list of the woods that we have lost, or may have lost, because of the ambiguity of some of the evidence. For example, in 1622 Sir John Grant of Freuchie obtained from the laird of Lundie in Glengarry rights over certain woods in Morar 'growand and standand upone thair landis of Killeismorarache, Kilnamuk, Swordelane, Arethomechanane and Brakegarrowneintoir'.[27] These may or may not have been pinewoods, and although the names are all unrecognisable now, the places to which they refer may or may not still be wooded under a different name. Similarly, place names containing the element 'ghuisachan' or some variant that are now bare of trees are not infallible evidence of a former pinewood. Some are easily misinterpreted. Carn a' Phrisghiubhais in Glen Einich on the Rothiemurchus estate, translated sometimes as 'hill of the

[25] S. House and C. Dingwall, '"A nation of planters": introducing the new trees, 1650–1900', in T. C. Smout (ed.), *People and Woods in Scotland* (Edinburgh, 2003), pp. 128–57.
[26] M. Stewart, 'Using the woods, 1600–1850: 2. Managing for profit', in T. C. Smout (ed.), *People and Woods in Scotland*, p. 126.
[27] National Archives of Scotland (NAS): G.D. 248/8. no. 342.

prized pine trees', actually means 'hill of the pine bush' (or clump). Others may refer to resources of bog pine at high altitudes, characteristically 4,000 years old and no evidence of any wood in historic time.

These problems apart, there is unambiguous evidence of about a dozen pinewoods obliterated (or nearly so) since 1600. Some of these were probably quite small. For example, in 1635 Sir James Campbell of Lawers sold 315 fir trees growing in the 'wood of Corriechurk', which was perhaps Corriecharmaig in Glen Lochay; the wood was never heard of before or since, but its disappearance is unlikely to have been solely due to this modest felling episode.[28] Others were undeniably large. Timothy Pont, *circa* 1590, depicted in his sketch map of Wester Ross a wood at least twelve miles long, stretching on both sides of the Gruinard river and its tributaries and clothing both banks of Loch na Sealga, with the comment at two points, 'fyrwood', and in an area north of Loch na Sealga he wrote 'mechtie parck of nature', a phrase redolent of open country with scattered trees, like a gentleman's hunting park.[29] The wood was still there at the time of Roy's military survey, *circa* 1750, but very much reduced, and of it today there is little more than a trace. Yet at no point in the past 400 years has there been any record of timber extraction. It seems that its disappearance could have been entirely natural.

Pont also placed a large wood along the southern shore of Little Loch Broom, without describing it further, and this was the area of which the Earl of Cromartie gave an account in a report to the Royal Society in 1710.[30] He described how, as a young man in 1651, he passed

> a Plain abut half a Mile round . . . all covered over with a firm standing Wood; which was so very Old, that not only the Trees had no green Leaves, but the Bark was totally thrown off; which the Old countrymen, who were in my Company, told me, was the universal manner in which Firr Woods did terminate.

In other words, he seems to have seen the very last traces of the wood that Pont depicted, and to have met people who were familiar with the phenomenon of woodland decay and regarded it as part of the natural order of things. Fifteen years later he passed that way again and found 'not so much

---

[28] NAS: R.D. 1/502, f. 121.
[29] Stone, *Pont Manuscripts*, p. 38.
[30] G. Mackenzie, 'An account of the mosses in Scotland', *Philosophical Transactions of the Royal Society*, 27 (1710), pp. 296–301.

as a Tree, or appearance of the Root of any: but in place thereof, the whole Bounds, where the Wood had stood, was all over a plain green ground, covered with a plain green Moss'. The countrymen told him the trunks had slipped away into the moss, 'which was occasion'd by the moisture that came down from the high Hill above it'. As at Loch na Sealga, there was no hint at all that human action had any part in its demise.

Other northern woods that vanished early without record of human cause include the Coille Mhór in Strath Nairn, now an almost bare rounded hill rising to 657 metres with a few scattered pines: the wood, according to local tradition, was destroyed to discourage outlaws on the orders of Mary Queen of Scots, but this unlikely explanation received no support whatever from historical records. Another is in the upper parts of Mar, in Glen Derry, described as full of huge collapsing pines 'some thirteen feet in girth' blown over by the wind, and in Glen Lui (probably Glen Luibeg) collapsing into the bog, in a way reminiscent of the Earl of Cromartie's account.[31]

Further to the south and along the west coast, pinewoods disappeared or declined that had certainly in several cases been the subject of human exploitation, sometimes heavily so. In the vicinity of Loch Hourn relatively little now remains of a wood enclosed by the Commissioners of Forfeited Estates in 1774 and apparently supporting a sawmill in the area as late as the Napoleonic Wars; a wood nearby in Knoydart, that Thomas Pennant reported seeing regenerating freely following a fire, has gone without a trace.[32] At Conaglen in Ardgour there are still relatively extensive woods of mixed pine and birch, but it does not seem to measure up to the account of around 1630 of 'Great number of firr trees in this glen . . . and there is a water in the glen which doeth transport great trees of firr and masts to the seasyde'.[33] More indubitably gone are the woods of Glencoe, the subject of an attractive map by Timothy Pont with the words 'Many Fyrre Woods heir alongs',[34] decorated with tree symbols that were probably added later by Robert Gordon of Straloch. The Glencoe woods were extensive and belonged partly to the Duke of Argyle and partly to Stewart of Appin; timber was sold from them to the Lowlands and to Ireland until at least the middle of the eighteenth century, though they had gone fifty years later. Similarly,

---

[31] Smout et al., Native Woodlands, p. 45; C. Cordiner, Antiquities and Scenery of the North of Scotland (Banff, 1780), pp. 27–30.

[32] H. M. Steven and A. Carlisle, The Native Pinewoods of Scotland (Edinburgh, 1959), p. 161; Thomas Pennant, A Tour in Scotland and Voyage to the Hebrides, 1772 (London, 1776), vol. I, p. 396.

[33] Smout et al., Native Woodlands, p. 104.

[34] Stone, Pont Manuscripts, p. 91.

there were three or four apparently small pinewoods on Loch Etive, now virtually gone, for which seventeenth- and eighteenth-century contracts exist showing that pine was exported from them.

Most striking of all is the case of the Glen Orchy pinewoods, their remnants known as the pinewoods of the Black Mount and the subject of an attractive recent book.[35] The pinewoods were sold to an Irish partnership in 1723, and greatly impressed the buyers for the scale and quality: one of them called them 'the best in Britain', which probably implies that many of them were close-grown and straight. At that time, they covered a considerable area in Glen Orchy and the surrounding glens, and the seller, the Earl of Breadalbane, specified that none should be cut that fell below two feet circumference at three foot from the ground. The Irish set to work but two years later the Earl arrived in person on a tour of inspection and found a scene of devastation – they had cut more than half the pine without leaving a standing tree, and were set to fell the remainder in the same way. He wanted to sue them for breach of contract, but then realised that he could find no trees below two feet in circumference in all the remaining woods: they were all 'very tall trees', and the Irish continued to fell them until 1736 at least.[36]

What had happened was that the Irish swept the forest clear of pines, except in a few inaccessible places, because there were few young ones in the wood. In other words, there had been a prior failure of regeneration, perhaps because the trees were so close-grown that they could not regenerate beneath the canopy and then because for some reason they could not do so by 'shifting their stances' to the moor beyond.

Taking all the evidence of the disappeared pinewoods in their entirety, one is struck by the fact that they all relate either to the west coast, or to woods at high altitudes. Furthermore, those cases where they disappeared without any known human agency are all concentrated in the north and north-west, while those on the west coast, south of Knoydart, were subject to felling episodes that removed the mature standing trees, possibly as in Glen Orchy in circumstances where there was little or no natural regeneration. Even today, regeneration occurs much more readily in the drier east coast areas than on the west, where moss and peaty ground make a less suitable seedbed. This suggests that an element of climatic stress is implicated, and that the woods were at least partly victims of the period of exceptionally cold and wet, windy, oceanic weather, associated with the later sixteenth and seventeenth centuries,

---

[35] P. Wormell, *Pinewoods of the Black Mount* (Skipton, 2003).
[36] Smout *et al.*, *Native Woodlands*, ch. 13.

and not completely finished before the nineteenth century. In these circumstances, pines in marginal places at the edge of their range were likely to die and regeneration to fail. It would happen first in the north and at great heights: in continental Europe, one effect of the worst decades of this period, felt from the Vosges to the Sudenten mountains, was to lower the treeline by 200m,[37] and a similar effect might be expected on the high ground and oceanic rim of Scotland. Where the climate was slightly more benign, as south of the Isle of Skye, mature trees might last longer but regeneration largely fails, and then ultimately becomes impossible when the mature trees are felled.

In prehistory, Scots pine had ebbed and flowed across the face of the country, showing, in Tipping's words, 'exceptional mobility and fragility because pine is very sensitive to climate change'. At one point, about 7,000 years ago, Galloway had had natural pine; in Caithness and Sutherland, less than 5,000 years ago, pine briefly reached the north coast and suddenly retreated again. On Rannoch Moor, pine disappeared less than 6,000 years ago, and in most places, fossil pines eroding out of the peat turn out to be 4,000 years old.[38] If climate change was a major factor in prehistory, it is also likely to have been so in recent centuries.

To look for a single explanation, however, would be to oversimplify, as there is also evidence that regeneration in the west could, on occasions at least, readily take place: as at Knoydart after fire noted by Pennant in 1772, and even in some Argyll glens if Smith in his evidence of 1798 really is speaking of Argyll and not upland Perthshire (one of his examples relates to the parish of Little Dunkeld). In these cases and others we should suspect any problems caused by climate change to have been exacerbated by overuse and overgrazing. What can be ruled out as a cause for total loss (except possibly in the case of Knoydart) is the arrival of commercial sheep farming, as the lost woods effectively disappeared before the sheep arrived. Others in the eighteenth century already of open structure, like Rhidorroch, were then grazed predominantly by cattle. That is not to deny that the coming of sheep made things worse in the nineteenth century in the manner described by Nairne.

In conclusion, the story of the lost woods may not in the last resort be one of human use and greedy insensitivity towards the environment, as it is

[37] H. H. Lamb, 'Climate and landscape in the British Isles', in S. J. R. Woodell (ed.), *The English Landscape: Past, Present and Future* (Oxford, 1985), pp. 148–67.

[38] R. Tipping, 'Living in the past: woods and people in prehistory to 1000 BC', in Smout (ed.), *People and Woods*, pp. 14–39; see also M. C. Bridge, B. A. Haggart and J. J. Lowe, 'The history and paleoclimatic significance of subfossil remains of *Pinus sylvestris* in blanket peats from Scotland', *Journal of Ecology*, 78 (1990), pp. 77–99; A. J. Gear and B. H. Huntley, 'Rapid changes in the range limits of Scots pine 4,000 years ago', *Science*, 251 (1991), pp. 544–7.

often portrayed, but of natural climate change as the primary cause and, in some cases, of human exploitation as a secondary one. It might even be considered that Scots pine, a continental tree in Scotland at the oceanic edge of its world range, has not naturally been very widely or continuously distributed here under the climatic conditions of the last 500 years, and that it was easily further discouraged by episodes of adverse climate change within that period. This would help to explain why so many well-intentioned schemes for pine planting or natural regeneration, financed by native woodland grants in the last decade in the uplands outwith the main strongholds of the species in Speyside and Deeside, are struggling or failing today.

Fig. 5 Craleckan iron furnace on Loch Fyne. Less well known and well preserved than Bonawe furnace on Loch Etive, it was also established by an English company, and operated from 1755 until 1813. The two between them called upon the resources of at least 20,000 acres of oak coppice when production was at its height, but as much wood was left in Argyll when they ceased operation as when they began. Photographer: John Hume. © Royal Commission on the Ancient and Historic Monuments of Scotland. Reproduced courtesy of J. R. Hume.

CHAPTER 5

The Atlantic Oakwoods as a Commercial Crop in the Eighteenth and Nineteenth Centuries*

The oakwoods of the western Highlands, and especially those that fringe the sea lochs of Argyll, are one of the glories of Scotland. In many parts of northern Europe and North America, fiords and inlets clothed with conifers are as frequent as they are beautiful. Much rarer is the sight of marine inlets clothed with oaks, and in fresh leaf in May they are breathtakingly beautiful. Ecologists call them Atlantic oakwoods, and are especially excited by the richness of their mosses and lichens that proliferate in the moist and mild climate, our native rainforests. Oakwoods are also common in a belt across Lochlomondside and the Trossachs and into Perthshire, but it is in Argyll that they are at their most distinctive and interesting.

The woods have always been an important resource to people, and we cannot entirely rule out the possibility that the earliest human settlers in Scotland were instrumental in their spread, as scientists are bewildered to understand how they advanced at such speed across the landscape of Britain.[1] From the beginning they were used, but they were not bought and sold. The market only became a big force in the Scottish Highlands in the eighteenth century.

The commercial exploitation of the Atlantic oakwoods involved the sale of the trees or their produce by the landed proprietors to capitalists, such as timber merchants, ironmasters or tanbark dealers. It did not normally involve sales by farmers, since they were tenants who did not own the trees and from about 1760 they usually had to buy timber themselves from the

---

* This is a revised version of a paper given at the Atlantic Oakwood symposium held by the Botanical Society of Scotland at Oban in 2005, and published in *Botanical Journal of Scotland*, 57 (2005), pp. 107–14. Thanks to the Society for permission to reproduce.
[1] S. Brewer *et al.*, 'Postglacial history of Atlantic oakwoods: context, dynamics and controlling factors', in symposium on Atlantic oakwoods, *Botanical Journal of Scotland*, 57 (2005), pp. 41–58; A. Lowe *et al.*, 'Route, speed and mode of oak postglacial colonisation across the British Isles', in ibid. pp. 59–82.

estates, or alternatively to use imported Norwegian timber, even in Argyll where they were surrounded by native timber. Most of the big purchasers were outsiders, that is, Lowlanders, Irishmen and Englishmen, who had the necessary access to capital and markets.

As a period in the total history of the Atlantic oakwoods, the episode of commercial exploitation was quite brief, in the main lasting from about 1700 to about 1900, and on a relatively intense scale only from about 1750 to about 1850. If we reckon that natural or semi-natural woods have been occupying the site for 8,000 years (and some have been here longer) that represents only 2.5 per cent and 1.25 per cent, respectively, of the total time-span that the woods have been with us. In their ecological history, it was a brief moment. As Sansum in particular has shown, the woods have always been highly dynamic.[2] They had plainly not remained in some unchanged pristine state for thousands of years before 1700, and with climate change in the offing, they are not likely to remain unchanged in the future. The precise ecological impact of the period of commercial exploitation is something to explore and remains uncertain in several ways, yet it tends to be uppermost in our minds because we live in the lee of this moment. It affects both our cultural and scientific perspective on the woods.

Commercial exploitation is not to be confused with anthropogenic impacts of an earlier type. People had been using the woods for thousands of years, since the Mesolithic, in varying ways and with varying intensity according to the weight of human numbers and their types of land-use practices. The woods had been a place to hunt, to gather fuel and food, to find building, tanning and dyeing materials and to feed and shelter stock. These activities are likely, by the end of the medieval period, to have been heavy and sustained enough to alter the character of the wood: Sansum suggests from palynological evidence that winter browsing by cattle and goats combined with regular cropping for timber led to declines in elm and ash in the woods, and probably in places to dwindling woodland cover.[3]

One other medieval use is worth picking out because it also had the potential to alter the woods in something of the same manner as commercial use: this is the cutting of oak, and indeed of pine, to make galleys for the chiefs. Some fleets of birlinns were comparatively large: the Lord of the Isles is said to have sailed to the Clyde for his wedding in 1350 with sixty galleys.

[2] P. Sansum, 'Historical resource use and ecological change in semi-natural woodland: western oakwoods in Argyll, Scotland', unpublished University of Stirling Ph.D. thesis, 2004; P. Sansum, 'Argyll oakwoods: use and ecological change, 1000 to 2000 AD – a palynological-historical investigation', in *Botanical Journal of Scotland*, 57 (2005), pp. 83–97.

[3] Sansum, 'Argyll oakwoods', p. 90. See also P. Quelch, 'Structure and utilisation of early oakwoods', in *Botanical Journal of Scotland*, 57 (2005), pp. 99–106.

Some of the boats themselves were large: Robert the Bruce demanded a galley of forty oars in return for a grant of lands in Argyll, and John of Lorn in 1354 got the right to build eight vessels of twelve to sixteen oars. Even as late as 1678, the Earls of Argyll could muster for an expedition against the Macleans of Duart, one galley (of twenty oars), four birlinns (of twelve to fourteen oars) and seventeen other boats including five scouts, probably with eight oars.[4]

These ships would demand trees large enough to be cut or cloven into planks. We know that such existed in places in the western Highlands, if only because of the description of the woods of Loch Maree around 1600 with 'great plentie of excellent great oakes, whair may be sawin out planks of four sum tyms five foot broad'.[5] But that such wood may have become difficult to obtain locally, at least for a time, is strongly suggested by a moment in Irish history in 1568 when the Earl of Argyll was forbidden access to Wicklow, Arklow or Dublin to get oak for building his galleys.[6] He would hardly have needed to come to Ireland had good ship timber been available on his doorstep.

Normally of course, the west coast lords did not buy timber, either from abroad or from one another, so shipbuilding was not a commercial operation but an example of feudal appropriation, obtained directly off their own lands or rendered in service as tribute from vassals.

There was a moment in the early seventeenth century, in 1611, when it seemed that the age of commerce had indeed broken into the western woods. Sir George Hay from Perthshire, in partnership with capitalists from the Sussex Weald, obtained from the Earl of Seaforth liberty to construct ironworks at Loch Maree, and to combine it with a timber-shipping business to exploit the local resource. Perhaps this finished off the great trees just described and lasted long enough to convert what remained of old growth in that part of Wester Ross into secondary woodland, but it was a flash in the pan, essentially finished by about 1630.[7]

A firmer start to commercialisation was made through the enterprise of the Irish, when they found that the boot was on the other foot from what it

---

[4] T. C. Smout, A. R. MacDonald and F. Watson, *A History of the Native Woodlands of Scotland* (Edinburgh, 2005), pp. 41–2; D. C. McWhannell, 'The galleys of Argyll', *The Mariners Mirror*, 88 (2002), pp. 14–32; D. C. McWhannell, 'Campbell of Breadalbane and Campbell of Argyll Boatbuilding Accounts 1600–1700', *Mariners Mirror*, 89 (2003), pp. 401–15.

[5] A. R. Mitchell (ed.), *Geographical Collections relating to Scotland made by Walter Macfarlane* (Scottish History Society, Edinburgh), vol. 2 (1906), pp. 539–40.

[6] R. Loeber, 'Settlers' utilisation of natural resources', in K. Hannigan and W. Nolan (eds), *Wicklow, History and Society: Interdisciplinary Essays on the History and Society of an Irish County* (Dublin, 1994),

[7] Smout *et al.*, *Native Woodlands*, pp. 229–36.

had been when they refused to sell the oaks of Wicklow. By the second half of the seventeenth century, they found that their own supplies of wood and bark could not meet the needs of a rapidly growing population, a vibrantly expanding capital at Dublin and a burgeoning tanning industry. From around 1670, they begin to turn up on the west coast of Scotland all the way from the Solway to the Great Glen, primarily in search of oak bark but also in search of pine where they can find it. In one case, at Achnacarry on Loch Arkaig, they were involved in establishing another early ironworks, though not on a large scale. So numerous were the Irish tanbarkers that a mercantilist tract on the eve of the Union of 1707 proposed banning them from Scottish shores, to encourage the Scots themselves to use the resource exclusively for tanning their own leather.[8]

In the eighteenth century, Irish ambition culminated in the extraordinary and ill-fated partnership led by Roger Murphey, a tanner variously described as of Dublin and Enniskillen, and Captain Arthur Galbraith, a retired military man, who with their friends began from 1721 to purchase woods in Argyll and further north. They signed at least eleven contracts and negotiated for two more, ranging all the way from Kennacraig to Letterewe. The largest were with the Earl of Breadalbane in Glenorchy and with Campbell of Lochnell, from whom they leased eight separate woods including those in Glen Kinglass and on the Isle of Eriska. They had some interest in pine and became infamous for the damage they inflicted on the woods of the Black Mount, for which they have been rather unfairly blamed – but that is another story. Their main focus was, certainly initially, on the oakwoods for tanbark and charcoal: their most positive contribution to industrial history was to establish a relatively sophisticated blast furnace and forge at Glen Kinglass in 1725, which lasted for a dozen years and was the direct forerunner of the better-known English operations at Bonawe and on Loch Fyne. The venture fell to pieces due to the partners' dishonesty and incompetence. Murphey was executed for murdering an employee, and the Irish faded from the scene in the 1730s.[9]

This left the scene to the English, who took over in respect of charcoaling and iron manufacture, from the 1750s, and to Scottish enterprise in respect of selling tanbark to leather manufacturers, particularly those in Glasgow and the Clyde area where the processing of cows' hides and the making of shoes was stimulated by consumer demand at home and export abroad to the American colonies.

[8] Ibid. pp. 200–1, 248–51; J. W. Burns (ed.), *Miscellaneous Writings of John Spreull* (Glasgow, 1882), pp. 63–4.
[9] Smout *et al.*, *Native Woodlands*, ch. 13.

Fell and Lindsay both gave authoritative accounts of what was involved in the two English iron-manufacturing enterprises, refuting the notion of Fraser Darling and others that the ironmasters bore a particular responsibility for the destruction of the woods.[10] The Bonawe furnace on Loch Etive was opened in 1753 by Richard Ford and Company of Furness in Lancashire, and Craleckan furnace on Loch Fyne in 1755 by Henry Kendal and Richard Latham and partners, also of Furness.[11] Production at Bonawe was of the order of 700 tons a year by the end of the century, and that at Craleckan probably similar. Bonawe made pig iron that was sent to Lancashire to be forged into a high quality product, but it also produced cast cannon balls as shot for the navy in the Napoleonic Wars. Craleckan had a forge on the spot. To keep them both going at this rate, assuming that oak trees were harvested for charcoal on a 20–24 year rotation, would have demanded the resources of 20,000–24,000 acres of wood. Contemporaries gave widely varying estimates at to how much oak there really was in Argyll around 1800, from 30,000 acres at the bottom end to 98,000 acres at the top, the latter probably including much scrubby pasture. Modern figures of ancient woodland are about 55,000 acres, and this may be close to what was there at that time.[12] Clearly, the demands of the two furnaces had the potential to affect a large proportion of the oak in the region, especially of those woods within reach of the sea. This can also be detected from the extent of the search for charcoal for Bonawe between 1786 to 1810, stretching all the way along the coast from Jura to beyond Ardnamurchan.

Craleckan lasted until only about 1813; Bonawe continued at a lower level of production until 1876, albeit with only intermittent activity after 1850. Probably the main reason for the greater longevity of Bonawe was a highly favourable lease that the owners signed with Campbell of Lochnell. This enabled them to buy wood for coaling at a fixed price for 110 years from 1754, which no-one could foresee would include the steep inflation of the Napoleonic Wars when such a lease was extremely advantageous to the company in securing an important part of their fuel supply. For the Craleckan company, when the Duke of Argyll was presented with an opportunity to renew a similar lease around 1811, he took the opportunity to increase the rent and the company shortly thereafter ceased operation.

By then, at least, the driver in the cutting and management of the oakwoods was no longer charcoal but tanbark markets. While the first spur

---

[10] A. Fell, *The Early Iron Industry of Furness and District* (Ulveston, 1908); J. M. Lindsay, 'Charcoal iron smelting and its fuel supply: the example of Lorn furnace, Argyllshire 1753–1867', *Journal of Historical Geography*, 1 (1994), pp. 283–98.

[11] G. P. Stell and D. H. Hay, *Bonawe Iron Furnace* (Edinburgh, 1995).

[12] Smout *et al.*, *Native Woodlands*, pp. 64–6.

towards better care of the woods had come from the intervention of the English ironmasters, by the end of the century, bark was usually worth twice the value of charcoal in a coppice rotation, and many owners considered a combination of tanbark and spokewood (used for wheels and ladders, etc.) sufficient justification for managing their woods for the market, irrespective of the ironworks. But the English were given credit where it was due and several commentators remarked that by giving a higher value to the woods they had encouraged owners to manage them better. This was the judicious summary of the Agricultural Reporter on Argyll for 1798: ever since their two ironworks had been set up,

> our natural woods are in general tolerably cared for: and though the long leases granted to those companies of some of the woods, and the want of sufficient compensation for the rest, has hitherto kept some of them low; yet they are always of more value to proprietors than any other equal extent of ground, arable land excepted.[13]

It was the opinion of Lindsay after careful investigation of Muckairn parish on the doorstep of Bonawe, that there is no evidence of the area under wood having fallen because of the ironworks during the time-span of their operation.[14] Indeed, it would have been economically irrational to allow it.

Although the extent of the woods did not contract, the impact of their commercial management on woodland structure was likely to have been considerable, though not differing much if the main outcome was to be charcoal or tanbark or both. Firstly, they were subject to systematic coppicing for the first time. Farmers in earlier periods had pollarded oaks and no doubt casually coppiced them, more or less at random, for their own uses. Now the woods and groves of an area would be systematically divided into haggs or felling coups and subjected to cropping on a rotation that varied from twenty to thirty years, but was often, as in the main woods supplying Bonawe, reckoned at twenty-four years. Each hagg would then be fenced to make it cattle-proof, and the 'spring' from each stool reduced to a smaller number of shoots than would occur naturally, to give stronger growth. Domestic animals were excluded for a minimum of five to seven years, though in the first half of the nineteenth century the exclusion period was sometimes lengthened to ten, fifteen or twenty years, and in some cases was total. But the last practice was regarded, for instance in Ardnamurchan in 1838, as bad management, as through 'the sacrifice of so much low land,

---

[13] J. Smith, *General View of the Agriculture of the County of Argyll* (Edinburgh, 1798), p. 129.
[14] Lindsay, 'Charcoal iron smelting'.

where wintering is much wanted'.[15] Possibly longer exclusions and longer rotations took place where bark and spokewood rather than bark and charcoal were the objective of production. Any exclusions at all were resented by the farmers whose animals needed winter feeding and shelter, and the minister of Muckairn in 1845 considered too much consideration was given to wood producing interests and too little to agricultural ones.[16] At least there was almost always a very serious effort made by the estate to integrate farming and silviculture. Grazing was in any case supervised, often by children, and damage to the trees would be noted and punished by the estate if it took place on a large scale.

As the coppice system spread, we may imagine the face of the countryside altering. There would soon be no old trees left, pollard or otherwise, except by accident or in difficult places. Within a generation it would be a world of coppice-with-standards, and the density of standards would be fairly sparse as they are really only kept for casual estate use. A quarter or a fifth or so of the wooded ground would be very young regrowth fenced from animals, so presumably full of butterflies and flowers in summer, and the remainder would be, to our eyes, a grazed pole-wood. As long as the system is maintained, the trees are in distinctive patches of unequal height, but they do not fruit or shed pollen. To the palynologist, they become invisible.

There were other possible effects too. Firstly, gaps and openings maintained by grazing animals in existing woods were often planted up with acorns, reducing their grove-like character and making each wood denser and more uniform. Some new oakwoods were also created, and when old stools died or appeared exhausted they were replaced in the same way. There is no reason to think that the foresters would favour acorns of local provenance, and some to think they preferred, if they could get them, English acorns, which may have impacted on the genetic character of the woods by introducing pedunculate oak or encouraging hybrids that might be less fertile.

Secondly, the foresters were advised by the experts to get rid of trees other than oaks as a waste of space, as due to its superior tannin content the value of oak bark was much greater than that of other barks. Robert Monteath in 1827, for example, told his landed readers: 'Oak, and nothing but oak, is the only profitable tree for coppice cutting, and whenever such a plan is intended, nothing else should be reared.'[17] Whether in practice the regeneration of birch, willow, rowan and hazel, or even of alder, elm and gean would

[15] Quoted in M. L. Anderson, *A History of Scottish Forestry* (Edinburgh, 1967), vol. 2, p. 100, from the *New Statistical Account* (Edinburgh, 1845), vol. 7.
[16] *New Statistical Account*, vol. 7, pp. 150, 520.
[17] R. Monteath, *Miscellaneous Reports on Woods and Plantations* (Dundee, 1827), p. 121.

have been excluded from the haggs once they were felled is another matter. Nevertheless, if Monteath's opinion was widely shared it would presumably have been acted upon by hacking down the saplings of unwanted trees and allowing their subsequent regrowth to be eaten down again by stock or to be crowded out by oaks that had been left.

The care taken in trying to establish an oak monoculture could be expected to vary from one site to another. My personal and unsystematic impression is that ancient woods in remote locations on the islands, such as Islay and Mull, which were probably only ever lightly managed, have a higher degree of tree-species diversity, and especially more non-oaks among the total trees, than those on the mainland of Argyll closer to the ironworks and the main tanbark centres, unless there is gorge woodland involved – but Sansum has clear indications that some of these mainland woods did also remain diverse.[18]

One can perhaps read too much into a single quotation, but I am struck by the difference between the present appearance of the woods round Inveraray and those described in 1751, on his farm, by Dugald Clerk of Braleckan in a letter to the Duke of Argyll following complaints that his men had damaged the Duke's oaks:

> The whole different species of woods upon my lands such as ash, elm, birch, hazel, alder, rowan, gean, quaking ash [aspen], sallies [willows], haw and sloethorn trees, etc. are disponed over to me by your Grace's predecessors and only the oaks . . . reserved for your grace.

But, he went on, in most places the oaks made up only a small part of the woods, and the other trees were:

> so intermixed and grow so thick and close together in the form of a hedge or one continued thicket, galling, rubbing upon and smoothing one another for want of due weeding and pruneing that it is morally impossible for any person to enter in to these thickets to cutt or weed either any of your Grace's oaks or my woods without at the same time breaking or cutting less or more of the other in order to get access to those thickets.[19]

It just does not sound like a modern Atlantic oakwood as we see them today, though in the trial plot at Glen Nant, enclosed since 1992, there is at least

[18] Sansum, 'Argyll oakwoods', pp. 90–3.
[19] National Library of Scotland MS. 17665, folio 66.

one comparable thicket, though whether the species mix is similar to that described above is hard to say.

A third possible modification is that too frequent coppicing reduced the nutrient condition, the timber yield and the floral diversity of the woods, as has been found to be the case in the rather similar woods at Coniston in the English Lake District, and in woodlands of the Sheffield region.[20] A similar effect was alleged (in respect of yield) by one contemporary in Perthshire.[21] On the whole, however, this is unlikely to have been effective on a wide scale in the western Scottish oakwoods, as systematic coppicing lasted in most places for barely a hundred years, and in much of the area for a shorter time, whereas in the English lakes and (to a lesser degree) in mid-Perthshire it took place over many centuries.

After the end of the Napoleonic Wars in 1815, the price of tanbark began to fall, and the returns from charcoal iron manufacture were undercut by increasingly efficient coal-fired blast furnaces, despite a quality premium remaining for charcoal iron. In neither case was the decline catastrophic in the short run. Bonawe was still producing 400 tons of iron in 1839, and in 1845 it was remarked that almost 600 people found employment at some seasons in the woods of Muckairn: most of them would have been women and children working on the bark. In the five years 1832–6, 400,000 trees were planted in Inveraray parish, of which 40 per cent were oaks put in to fill the gaps and to replace exhausted coppice stools in the natural woods.[22]

The real crisis began in the course of the 1840s with the precipitous decline in bark prices, £18 a ton around 1841, barely £6 a ton a decade later, and as little as £4 a ton by 1903. The root cause was foreign competition, particularly from Denmark, along with the advent of chemical substitutes for oakbark tannin. Yet the bleakness of the future was not immediately apparent to woodland owners. It only became plain in the 1860s that traditional coppice management would no longer pay.[23]

In 1881, Thomas Wilkie, the forester at Ardkinglas, reflected on the changed world. Forty years ago the return per acre in Inveraray parish had been eight times what it was now: the woods now gave better profit if left unenclosed and exposed to the full weight of grazing and sheltering stock.

[20] S. Baker, 'The history of the Coniston woodlands, Cumbria, UK', in K. J. Kirby and C. Watkins (eds), *The Ecological History of European Forests* (Wallingford, 1998), pp. 167–84; I. D. Rotherham, 'The implications of perceptions and cultural knowledge loss for the management of wooded landscapes: a UK case-study', *Forest Ecology and Management*, 49 (2007), pp. 100–15.

[21] *Old Statistical Account*, vol. 12, pp. 231–2.

[22] *New Statistical Account*, vol. 7, pp. 13–14.

[23] Smout *et al.*, *Native Woodlands*, ch. 10.

If a coppice was cut it was best to let the animals in at once, while reserving the old standards and fencing off a few new ones to develop from the old stools. The trees stood 350 to the acre, but should times change it would be easy to reconvert to coppice: if they did not, more of the standards would progressively be cut out as the wood was converted first to wood pasture, and finally, after 19 years, it would simply become pasture. Many advocated just singling the old stools, or praised the virtues of larch as a timber crop in place of the oak, or advocated (as Wilkie did in 1889) under-planting abandoned outgrown coppice with game cover such as privet, current, barberry, laurel, hazel, rhododendron, cotoneaster, sweet gale, gaultheria, elder and a few silver firs.[24]

Were there alternative uses to pasture and game? If there was a pyroligneous acid works nearby, as at Balmaha on Loch Lomondside, this perpetuated the coppice system to the end of the century. The Prussian forester Adam Schwappach, in 1896, observed 'numerous woods of oak coppice', admittedly badly managed but providing Balmaha with a thousand tons or more of smallwood a year.[25] But on the Atlantic coast there were no such factories. Oak could still be used as spokewood; as late as 1893 Scottish spokewood commanded a premium over American imports. Thirteen years later, however, the Americans had the edge, many wheels were imported prefabricated, and railway wagons had gone over mainly to American oak because it was larger. Native oak spokewood was still used in colliers' and contractors' carts, and as cross pieces in telegraph poles, but that was a small and declining market. The commentator hoped that manufacturing oaken spokes for the wheels of the new motor cars might provide a new market. Of course it was not to be. After two centuries, around 1900, the commercial age of the woods ended.

Those of the Atlantic oakwoods that did not succumb to conifer planting, reverted to grazing without periods of enclosure. They grew up to even canopies of uniform age, and they flowered and fruited again. They did not regenerate, perhaps because of the grazing, possibly because of genetic changes, possibly even because of new diseases such as that caused by American mildew, possibly because the jay, that great transporter of acorns, had largely disappeared. They still bear the mark of the commercial period, but they are high forest again, quite different in appearance from

---

[24] T. Wilkie, 'On the system of oak coppice management recently adopted', *Transactions of the Scottish Arboricultural Society*, 9 (1881), pp. 270–2; T. Wilkie, 'Report upon the rearing of underwood for game coverts in high forest', *Transactions of the Royal Scottish Arboricultural Society*, 12 (1889), pp. 371–3.

[25] A. Schwappach, 'Report of a visit to the forests of Scotland in August, 1896', *Transactions of the Royal Scottish Arboricultural Society*, 15 (1898), pp. 11–21.

what they were. But change has always been a hallmark of these dynamic woods, and with different objectives in their management, and external circumstances set to alter with climate change, they are bound to continue to change.

Fig. 6 Crofters on Harris working their peat bank, 1937. In the late 1940s, out of 153 households sampled in Lewis and Harris, 125 still used only peat for cooking, and the rest used peat and coal (Fraser Darling, *West Highland Survey*, p. 364). But in most of Scotland peat bogs were synonymous with wasteland, ripe for drainage and afforestation with Sitka. Photographer: R. M. Adam. Courtesy of St Andrews University Library: RMA-H5593.

CHAPTER 6

Bogs and People in Scotland Since 1600*

Peatland is abundant with us: there may be as much as 1.6 million hectares of the resource in the United Kingdom. Some of this, like the fens of Cambridgeshire and Yorkshire, have over the centuries been largely drained and turned into good agricultural land. Much, however, remains wetland. Blanket mire (also known as blanket bog) covers about a million hectares of Scotland. Raised bog, those remarkable domed structures of peat and sphagnum moss that receive all their moisture directly from the air rather than from streams running into them, cover a much smaller area, only about 27,000 hectares in Scotland.[1]

Of all this, little is unmodified by man, though much can still be classified as semi-natural. Only 9 per cent of the Scottish raised bogs and 11 per cent of the total UK mires even approach a pristine state.[2] Cut, drained, planted, bulldozed away, mires, fens and bogs have attracted the attention of people from time immemorial. In prehistoric times they were places of refuge and of sacrifice, where offerings of human victims and bronze wealth could be flung. In the Middle Ages they were clearly valued for their supply of fuel and building material, as monastic charters testify. One such was a confirmation by Robert the Bruce to the monks of Lindores in Fife of the lands of 'Monks-moss', with the right to carry off annually two hundred cart-loads of heather and all the peats from the 'peteria' that they might need.[3] When evidence becomes fuller, in Scotland from the seventeenth century onwards, the reputation of bogs as a useful resource stood high.

The first detailed geographical accounts of Scotland began to be written mostly between 1580 and 1730. At Fetteresso in Kincardineshire the parish was said to be supplied with 'inexhaustible mosses, wherein are digged the best of peats, very little if anything inferior to coals', from which the

---

* This is a rewritten and much expanded version of a paper of the same title in L. Parkman, R. E. Stoneman and H. A. P. Ingram (eds), *Conserving Peatlands* (Wallingford, 1997), pp. 162–7. Thanks to CAB International publishers for permission for reuse.
[1] L. Parkyn, R. E. Stoneman and H. A. P. Ingram (eds), *Conserving Peatlands* (Wallingford, 1997), esp. pp. 5–9, 188–91, 205, 425.
[2] Ibid. pp. 191, 425.
[3] R. Sibbald, *The History Ancient and Modern of the Sheriffdoms of Fife and Kinross* (Cupar, 1803), p. 385.

inhabitants supplied not only themselves and the neighbouring communities of Dunnottar, Inverbervie and Stonehaven, but also Aberdeen some fifteen miles away.[4] At Cortachy in Angus, 'the hills and glens of this county abound with excellent moss and muir for feuell, with wild fowl of different kinds, and sometimes with deer and roe'.[5] A description of Aberdeen and Banff by Robert Gordon of Straloch, probably of the 1630s, says that 'there is no occasion here for stoves; the hearths are well supplied with peat, which is dug out of the ground, and is black and bituminous, not light and spongy, but heavy and firm'.[6] A parish with 'moss ground' was blessed, like Keith: it had 'great plenty of fir under ground, which the people thereabouts dig up some two fathoms deep, and by this they are served with winter light and timber for their houses. In this hill is a large peat bank about six or seven foot deep and near two miles long'.[7] A parish without such resources was cursed: of Cushnie it was said 'it is a poor countrey both for corn and pasture, and exceeding scarce of Fewel'.[8] Of the lower ground of Morayshire near the coast it was observed that 'they suffer from scarcity of peats for fuel, which is the only inconvenience felt by this highly favoured region, but even that in few places, and they remedy it by hard drinking in company, for this also must be admitted'.[9] If you could not warm yourself with peat, you needed to warm yourself with whisky.

This generally cheerful and positive attitude towards bogs may be contrasted with the attitudes that came to rule in the following century, the age of the agricultural Improvers, when the old assumption that natural resources were a given changed to one where they were regarded as generally capable of betterment. One of the first to consider Scottish bogs rationally and with a view to change of use was the Fifer, Sir Robert Sibbald, writing in 1710 a few pages of digression in his *History of Fife and Kinross* 'to give some account of the rise of the moors, mosses and bogs, and how they may be improved to better value'. He believed (erroneously of course) that they had been formed by the Romans felling the forests and thereby damming up the streams so that vegetation rotted, and then erosion by water and wind brought down more earth into the putrefying mixture. Now 'they increase annually, by the new grass and sedge growing upon the rotting of the old' (this was correct) and 'the moss grows not above an inch or so in a year's time' (an optimistic

[4] A. Mitchell (ed.), *Geographical Collections Relating to Scotland made by Walter Macfarlane* (Scottish History Society, Edinburgh, 1906), vol. 1, p. 248.
[5] Ibid. p. 284.
[6] Ibid. p. 268.
[7] Ibid. p. 89.
[8] Ibid. p. 31.
[9] Ibid. p. 457.

estimate). He identified mosses of different colours, white, grey and black, the latter the commonest and the best fuel. He surmised it to be 'a perfect putrefaction of the plants which grow upon these grounds'. For him, however, the important question was 'how these mosses may be converted to useful and profitable ground', either by draining 'where they are very soft and full of water', as had recently been done by Sir William Bruce at the edge of Loch Leven, to make 'good meadow and firm ground' or 'where the moss is not so soft and waterish' by 'burning it in a drouthy and dry summer', as his friend Lord Rankeillor had done outside Cupar, 'and made good arable and pasture ground'.[10] He noted that reclamation had also taken place in the carselands of Stirling, and in 1724 it was observed that parts of Flanders Moss 'by casting, pareing and burning' had been in some places 'cut quite thorow and made arable ground'.[11]

This was just the start of a national effort to encourage reclamation, comparable in some respects to the drainage of the Yorkshire and Cambridgeshire fens in England in the seventeenth century, though the technology was very different and those peatlands were not, in the strict sense, like the Scottish bogs in configuration or type. To the late eighteenth and early nineteenth century Improvers, peat bogs cried out for money and effort to transform them from waste into arable land, even though in parts of the countryside the exhaustion of peat supplies was already said in the 1790s to be leading locally to depopulation. Reports of this came from several counties in the 1790s, including Aberdeen, Dumfries, Inverness and Peebles.[12]

So efforts at reclamation met with acclaim. The great scheme at Blair Drummond, adjacent to Flanders Moss, initiated by the notable Improver Lord Kames but actually carried through by his heir, was hailed as a national benefit, greater even than that conveyed by David Dale in founding his famous cotton manufactory at New Lanark. Both had employed displaced Highlanders, wrote William Aiton in 1811, but whereas at Blair Drummond 'the moss colony remain healthy and happy, delighting in their situations, warmly attached to their patron, and to the Government, daily increasing in wealth, and rearing a numerous offspring, ready to extend their brawny arms, in the cultivation of the dreary wastes, or to repel their country's foes', those in the cotton mill 'became discontented with their situation, and soon abandoned it'. That 'several hundreds of ignorant and indolent Highlanders', went on Aiton, were 'converted into active, industrious, and virtuous cultivators, and many hundreds of acres of the least

[10] Sibbald, *Fife and Kinross*, pp. 153–6.
[11] Mitchell, *Collections*, p. 341.
[12] H. Hamilton, *Economic History of Scotland in the Eighteenth Century* (Oxford, 1963), p. 207.

possible value rendered equal to the best land in Scotland are matters of the highest national interest, to which I can discover no parallel in the cotton mill colony'.[13]

These tones were quite characteristic of early nineteenth-century commentators. Thus the Rev. Robert Rennie of Kilsyth, a pioneer in the systematic study of bog formation, wrote in 1807 that 'innumerable millions of acres lie as a useless waste, nay, a nuisance to these nations. The benefits that might accrue to Europe by a slight attention to this subject, are above all calculation. It is impossible for numbers to express, or the imagination to conceive, correctly, the extent of these'.[14] Aiton himself believed that 'the intrusion of Moss earth has been attended with two evils of great magnitude; first, the loss, or at least the reduction, of the value of an immense extent of soil, and secondly, its pernicious effects on the atmosphere'.[15] He estimated the amount of ground under bog in Scotland at over 14,000,000 acres, and believed that the accumulation of moss over so much of the original soil since (he thought) the time of the Roman invasion, had led to a decline in the temperature. Andrew Steele, in 1826, in the opening chapter to his treatise on peat moss, spoke of the bogs as 'immense deserts . . . a blot upon the beauty, and a derision to the agriculture of the British Isles'. He also commented that 'the only animals found on these grounds are a few grouse, lizards, and serpents'.[16]

The agricultural experts received general support in their view of bogs as dreary encumbrances from the ever-growing band of Romantic tourists, who came to Scotland to view the glens and to obtain a frisson of excitement from the 'picturesque', 'the sublime' or the 'terrific', but who found nothing attractive in bogs as they floundered through them. Thus, John MacCulloch, Walter Scott's friend, had a memorable diatribe against the Moor of Rannoch:

> Hideous, interminable . . . a huge and dreary Serbonian bog, a desert of blackness and vacuity and solitude and death . . . an ocean of blackness and bogs, a world before chaos; not so good as chaos . . . even the crow shunned it . . . if there was a blade of grass anywhere it was concealed by the dark stems of the black, black muddy sedges, and by the yellow melancholy rush of the bogs.[17]

---

[13] W. Aiton, *A Treatise on the Origin, Qualities and Cultivation of Moss-Earth, with Directions for Converting it into Manure* (Ayr, 1811), pp. 341–2.

[14] R. Rennie, *Essays on the Natural History and Origin of Peat Moss* (Edinburgh, 1807: also 1810), p. 6.

[15] Aiton, *Treatise*.

[16] A. Steele, *The Natural and Agricultural History of Peat-moss on Turf-bog* (Edinburgh, 1826).

[17] J. MacCulloch, *The Highlands and Western Isles of Scotland* (London, 1824), vol. 1, pp. 17–20.

When, however, the train replaced the pony as a way of crossing the bog, the traveller could view them more dispassionately and in greater comfort. The first thoroughly appreciative description of the aesthetics of the Moor of Rannoch came from the anonymous author of a public relations book for the new line. The moor becomes in winter 'twenty square miles of a study in sepia', in summer 'one colossal Turkey carpet, so rich and oriental'. Even: 'to see a sunset on Rannoch Moor is as essential as to see Loch Lomond by moonlight'.[18]

Nevertheless, until far into the second half of the twentieth century, the idea that the peat bog was a desert that needed reclamation, or at least was a wasted resource that could legitimately be used up for some economically productive purpose, ruled most thinking on the matter.

One reason was the declining usefulness of peat bogs to communities, as coal gradually became the fuel of choice even in areas some distance from the mines. An impetus was given to this in 1793 by the abolition of coastwise duties on Scottish coal: up to this point, even Aberdeen had still been spending £3,000–4,000 a year on peat brought into the city.[19] Rural areas, however, were only slowly weaned from peat. As late as 1843, when only one railway existed in Scotland (from Glasgow to Edinburgh, with a few feeder lines), and steam shipping was not a generation old, coal was the exclusive fuel only in the central belt and in Berwickshire. In the north, transport costs doubled or tripled the pit-head price, and in inland parishes it could rise six- or seven-fold. Parishes beyond the centre frequently still depended wholly or partially on peat.[20] A century later, after the spread of railways, the steam puffer and latterly the motor lorry, the situation was transformed. Peat was now used only in the remoter parts of the Highlands: Fraser Darling in the later 1940s found that 30 per cent of crofting townships were still dependent solely on peat for fuel, and 30 per cent dependent solely on coal (the remainder used two fuels) – at the extreme, in the Hebrides, 75 per cent of townships still depended entirely on peat, but 20 per cent also used coal.[21] The coming of the national grid, and Thomas Johnston's insistence, as head of the first Hydro Board after 1945, that electricity should reach the remotest areas, further emancipated even the crofting population from slavery to the peat spade.

As the economic importance of the bogs to local communities declined, so the call for alternative uses of the peatland 'wastes' became louder. Given

---

[18] Anon., *Mountain, Moor and Loch, Illustrated by Pen and Pencil on the Route of the West Highland Railway* (London, 1894), pp. 101, 107.

[19] Hamilton, *Economic History*, p. 267.

[20] I. Levitt and C. Smout, *The State of the Scottish Working-Class in 1843* (Edinburgh, 1979), pp. 56–7, 69.

[21] F. Fraser Darling, *West Highland Survey* (Oxford, 1955), pp. 301, 365.

the ploughing and drainage technology available in the nineteenth century, what could be done for agriculture was limited, but forestry seemed a slightly better proposition. Even in the early eighteenth century, attempts had been made to plant trees on Blair Bog in Peebleshire and at the Whim in Midlothian, and in the later nineteenth century more successfully at Loch Tulla, Argyll, Durris, Kincardineshire and elsewhere. The most famous and significant attempts were those by Sir John Stirling-Maxwell on the blanket bog at Corrour, Inverness-shire, begun in 1892, using Sitka and Scots pine and utilising the 'Belgian system' of digging ditches and planting trees on top of upturned turves. It was still limited in what could be achieved, and when the early Forestry Commission came into being in 1919 (with Stirling-Maxwell as its third chairman), it continued experiments on planting peat-lands at Inverliever, Argyll: 'it was the better quality peats on which the early planters were successful, not the nutritionally poor ones'.[22] The whole tenor of the forestry lobby's approach before and after the First World War was that much of upland Scotland was just wasteland, a reproach and an oppor-tunity for those who wished to see 'our bare glens re-peopled with trees and busy with the healthiest of industries'.[23] But foresters had to acknowledge that relatively little could yet be done on deep peats.

After the Second World War, the situation changed dramatically in several respects. First, technology made deep ploughing and draining a possibility on the hills, with the coming of the Cuthbertson plough in the 1950s and the discovery of how to fertilise Sitka and lodgepole pine to make it grow in the most unpromising places: this was matched for fen and field drainage by a new generation of drag-lines, bulldozers and mole ploughs. The area of Scotland under trees increased from 6 per cent of the land surface in 1960 to 17 per cent by the end of the century and most of that was Sitka planted on cheap peaty soils, often on deep blanket bog or raised bog. At first most of the effort was supplied by the Forestry Commission itself, using government money and striving for planting targets inflated in the 1960s by the demands of the Secretary of State for Scotland, who saw forestry as a solution to the depopulation of the Highlands. Later, it was private forestry companies, encouraged by tax breaks, who increasingly planted most of the ground, until Nigel Lawson's budget of 1988 pulled the rug from under the industry.

Agricultural subsidies also encouraged the transformation of many peatlands into farmed pasture. In the post-war decades, the call to reclaim

---

[22] A. R. Anderson, 'Forestry and peatlands', in Parkyn et al. (eds), Conserving Peatlands, pp. 237–8.

[23] Stirling-Maxwell, 'The planting of high moorland', Transactions of the Royal Scottish Arboricultural Society, 20 (1906), 1–7, quoted in Anderson, 'Forestry', p. 238.

marginal land of all descriptions reached a peak of frenetic zeal quite analogous to that of the early nineteenth-century Improvers. In a pamphlet entitled *Reclamation!* the Scottish Peat and Land Development Association (c. 1962) called for a Land Development Board to encourage the transformation of the waste: 'we can no longer afford to have so much marginal land put to so little use, or deteriorating through misuse'. Among its proposals were to establish experimental farms on bogs, and to initiate reclamation and improvement by concentrating drainage machines in groups 'in an area that forms a natural entity – e.g. the whole of a glen or a major bog'. J. M. Bannerman, the well-known Liberal, called for 'widespread arterial drainage schemes', especially in Strathspey 'where the land to be reclaimed runs into scores of thousands of acres'. He also proposed lowering Loch Lomond by four feet by removing the silted sandbanks at the outflow. Others called for an onslaught on the 4 million acres of partially productive land by 'chemical ploughing' (i.e. by massive application of herbicide). These people were visionaries but were not cranks: apart from Bannerman, they included respected MPs like Tom Fraser and the future Conservative Secretary of State for Scotland, Michael Noble.

Calls for large-scale drainage were fortunately not always heeded. In 1950, the recommendations of the Duncan Committee for a subsidy from the state to drain completely Lochar Moss in Dumfriesshire and the upper parts of Strathspey were rejected by government and, in 1990, calls for major engineering on the Spey were also turned down after a public enquiry, partly on the grounds that it would threaten the Insh marshes, recently reflooded upstream by the RSPB. Probably the most useful outcome of these years was a national survey of peat bogs begun by the Scottish Office after 1947 and taken over by the Macaulay Institute in Aberdeen. Although intended as the basis for commercial exploitation of the resource, its greatest utility in the end has been to aid nature conservation in accurate assessment of the bogs.

The ecological movement of the second half of the twentieth century was long ambivalent about the peatlands. W. H. Pearsall in his New Naturalist *Mountains and Moorlands* (London, 1950) wrote a text of classical importance about their geology, biodiversity and ecology, but concluded that in a small, crowded island there was tension between keeping the open upland spaces for recreation or for national production, 'a role which is becoming a necessity as a result of increasing population and diminishing trade balances'. He did not even mention nature conservation, but asked, since 'what we see in the uplands today is often the degenerating remains of the former plant and animal communities', why change should be considered objectionable. Both afforestation and improved pasture were acceptable in their place, he believed, but peat presented a special problem: it could make a productive soil, but it must be drained and limed to become so, and this was difficult and expensive. 'Thus often the only reasonable way of handling it is to try to drain it and to grow on it *mor* plants like pines'. In the end, he believed, it might prove easiest to keep the bogs for their 'great virtue not often realised', that of storing and releasing water.[24]

A similar tone was taken in a purely Scottish context by Fraser Darling and Morton Boyd in their 1964 revision of *Natural History in the Highland and Islands*; there is much on upland and insular ecosystems, though barely more than a page on bogs *per se*. When they talked of peat it was especially in the context of recent attempts to exploit it (these included an experimental peat-fired power station in Caithness, which unlike those in Ireland, was not a success). They note that 'one of the encouraging advances has been the development of peat lands for agriculture and forestry'. On Lewis, where the islanders had already over the centuries created 'skinned land' by removing the overburden of peat for fuel and fertiliser, they observed that modern tools like scrapers and bulldozers could help the agricultural scientist create much good land, and commended 'the heavy dressing of the peat cuttings and bog surfaces with lime, compound fertiliser and seeds of grasses and clover'.[25] Darling, like Pearsall, was anxious to portray a 'devastated countryside' and coined the phrase 'wet desert' to describe much of the Highlands.[26] Around 1973, he wrote to the forester H. A. Maxwell describing Sitka spruce as a 'godsend' on peaty ground that would 'recreate a forest biome after a long period of soil degradation'.[27]

All this was to change dramatically when a new generation of ecologists began to accept bogs for what they were, not for what they might have been or ought to be. In 1977, the great modern Domesday book of British sites, Derek Ratcliffe (ed.) *A Nature Conservation Review*,[28] listed 107 peatlands of which 37 were in Scotland, including most of the largest: Rannoch Moor was over 10,000 hectares. The message, unequivocally, was that these were of many different kinds, but all very precious and exceptional resources for nature conservation, and they were also of 'enormous value as historical

---

[24] W. H. Pearsall, *Mountains and Moorlands* (London, 1950), pp. 270, 281.

[25] F. Fraser Darling and J. Morton Boyd, *The Highlands and Islands* (London, 1964).

[26] Darling, *West Highland Survey*, esp. p. 192; F. Fraser Darling, *Pelican in the Wilderness* (London, 1956), p. 180.

[27] H. A. Maxwell, 'Coniferous plantations', in J. Tivy (ed.), *The Organic Resources of Scotland: Their Nature and Evaluation* (Edinburgh, 1973).

[28] D. A. Ratcliffe (ed.), *A Nature Conservation Review* (Cambridge, 1977), vol. 1, pp. 249–87; vol. 2, pp. 206–44.

records of conditions, both physical and biological', as exemplified by the palynological and allied studies of Harry Godwin.[29]

From there, events moved rapidly towards the confrontations of the 1980s. On one side was the forestry interest, especially private forestry companies pursuing the advantages of tax breaks for their clients, glad to buy cheap peatlands, drain them and plant them up with Sitka spruce and lodgepole pine. They were warmly backed by the Highland Council, who saw forestry as a means of reviving the rural economy, though in this they were to be entirely disappointed. On the other side were the increasingly exasperated and hostile conservation bodies, especially the Scottish Wildlife Trust and the RSPB, calling on the Nature Conservancy Council (with its headquarters in Peterborough in England) to take a firmer line in response to the threat to biodiversity. It came to a head in a row over the vast blanket bogs of the Flow Country in Caithness and Sutherland, most of it still not confirmed or reconfirmed as SSSI following the 1981 Wildlife and Countryside Act. The foresters believed the demands of the conservationists were unreasonable, and tainted by coming from outsiders. The conservationists believed the foresters were vandals, profiteers and outsiders too. Government became irritated. The Scottish Office, traditionally pro-afforestation, resented having to deal over a matter they regarded as critical to the rural economy of remote areas with a regulatory body in Peterborough. The Minister of the Environment at Westminster, Norman Ridley, under whose auspices the NCC operated, was equally resentful of wasting time dealing with a bad tempered quarrel in the north of Scotland. Malcolm Rifkind, the Scottish Secretary, in a rough and ready judgement of Solomon, declared half the Flow Country would be available to forestry, and half would be protected for the birds.

That satisfied no one. The chairman of the NCC declared the Flows to be the natural heritage equivalent to the Taj Mahal, but he did so at a press conference in London, six hundred miles from the problem. It was said in the north that the only Taj Mahal known in Sutherland and Caithness was an Indian restaurant in Thurso.

The conservationists, however, found an unanticipated ally in the *Daily Telegraph* and its environment correspondent Charles Clover, who argued that it was inappropriate in Mrs Thatcher's Britain for government to take the side of a small vested interest and support it with generous tax concessions while it wrecked a national ecological treasure. The government had just destroyed the miners and ignored their argument that coal would have

[29] *Nature Conservation Review*, vol. 2, p. 271; H. Godwin, *The History of the British Flora: a Factual Basis for Phytogeography* (Cambridge, 1956).

to be imported if it was not mined at home: a main argument of Fountain Forestry and the other private companies was similarly that Britain needed to grow her own wood and not to be so dependent on foreign supplies. Arguments about saving foreign exchange by import substitution might have worked in the past, but for a government committed to globalisation their time was over. In 1988, the Chancellor Nigel Lawson abolished tax breaks for forestry, and planting on the peat bogs ceased overnight. In due course, the RSPB bought some of the planted Flows back and began to restore them to bog land.

At the same time, Nicholas Ridley and Malcolm Rifkind agreed that the responsibility for nature conservation should be divided into separate bodies for England, Scotland and Wales to give each greater and more direct administrative control over nature conservancy. Scottish Natural Heritage and English Nature (and the Countryside Council for Wales) were born out of the dismembered NCC. In due course, they were each called upon to save more bogs from destruction, now generally by peat extraction for the horticultural industry. SNH saved the biggest raised bog in Britain, Flanders Moss, by buying it from a firm that had longstanding rights to destroy it. As it cost over £1,000,000 to do this, the purchase needed permission from Michael Forsyth, the then Scottish Secretary in whose constituency it also happened to be. He was appalled at the cost and puzzled by the necessity; SNH talked him round, but baulked at the prospect of approaching him again for similar sums for other bogs. The restoration of Flanders Moss then proceeded apace and must be counted one of the notable successes of the organisation's first decade. At Thorne Moors in Yorkshire, English Nature came to a much less satisfactory compromise with the developer, but is now also involved in restoration. Along with the RSPB work in the Flows, and the campaigns of the Scottish Wildlife Trust, much has been done to reverse the older perception of bogs as useless waste. They are certainly not yet entirely safe from further destruction (for example by opencast mining in the Central Belt of Scotland), but they are a great deal safer than they were.

So far we have considered how outsiders and largely self-defined 'experts' regarded bogs. We also need to touch briefly upon the rather more obscure topic of how those who had bogs upon their land actually managed them on a day-to-day basis. Peatlands and bogs were, of course, of many kinds: for practical use the most important variation was the spectrum from wet to dry. Most occupiers saw them as a source for fodder, building materials, fertiliser and (above all) fuel. Grazing maintained them as open country, free from trees: a comparatively light stocking can prevent regeneration, so that 'the past absence of dense forest in the uplands below the altitudinal limit of trees may well have been caused by the herbivores present', deer and cattle as well as sheep.[30] Domestic animals grazed the bogs and peatlands when they were accessible in the summer months, but pierhaps they were especially useful as sources of hay, and of spring bite.

This has to be seen in the context of earlier grazing regimes, where the main constraint on stocking levels was the ability to keep animals alive between November and May without artificial foodstuffs, turnips, or silage. Until well into the nineteenth century, explains John Mitchell, writing about Loch Lomondside and surrounding areas, 'cattle being over-wintered were fed almost exclusively on "bog-hay", a mixture of wet meadowland plants scythed from undrained land': it could yield between 100 and 150 stone of hay per acre, and some of the bog hay meadows were of great extent. What was said to be the largest in Scotland was the Carron Bog, four miles long and a minimum of one mile wide: it almost all now lies beneath the Carron Reservoir, but in its heyday in the late eighteenth century was described by the local minister as adding 'great liveliness and beauty to the general face of the country. The scene it exhibits during the months of July and August, of twenty or thirty different groups of people employed in haymaking, is certainly very cheerful'. You may still see bog hay cut today in County Donegal and Sligo, and no doubt elsewhere. The Aber Bogs on Loch Lomondside were last so cut in 1952, but the practice must now be unfamiliar at least in mainland Scotland.[31] On the other hand, most upland farmers still know of the value of bogs for spring bite, when the emerging sprouts of bog cotton and other plants provide a richness of early protein that is still welcome for animals coming out of winter quarters.

The use of the peat bog for opencast extraction by the occupier was obviously important, but it is significant that in some societies – Ireland, Denmark – the generic term 'turf' (qualified in different ways) is used for anything cut from the ground. The Danes distinguish between 'grass-turf', 'heather-turf', 'bog-turf' (or peat) and 'sand-turf' (peat under sand).[32] The point is that all were obtained by skinning the ground, and the difference to the farmer was in degree rather than kind. They were all suitable for cutting sods for burning, and subsequent use as fertiliser, a practice that in Scotland was viewed sometimes with favour, and sometimes with hostility by experts, but – at least until the middle of the nineteenth century – was widely practised by farmers. Turf, or 'grass-turf' to the Danes, was also used

---

[30] D. Welch, 'The impact of large herbivores on North British peatlands', in Parkyn et al., *Conserving Peatlands*.

[31] Mitchell, 'A Scottish bog-hay meadow', *Scottish Wildlife*, 20 (1984), pp. 15–17.

[32] T. T. Hove, *Tørvegravning i Danmark* (Herning, 1983).

as a building material, for 'feall dykes' (the Scottish term) and for 'divots', or scale-shaped roofing turves.[33] Certain qualities of peat could also be used for building, but it would have a much shorter life, particularly in frost: it would therefore normally be avoided except for buildings that were essentially temporary, like shieling huts. During the reclamation of Blair Drummond Moss, however, the homes of the Highlanders who were doing the work were made out of peat blocks.

The most significant use of peat, however, was for fuel, though in districts where peat was scarce – as in parts of Perthshire – turf was used in its place: the disadvantage of 'grass-turf' was that it burnt too fast. Even good quality peat was needed in very large quantities. Fraser Darling reckoned that a family in the western Highlands needed to cut 15,000–18,000 peats a year for cooking and warmth: a good man could cut 1,000 a day, but up to a month's work for the family was involved in winning peats, drying and carrying them.[34] (See also Chapter 7 for more on fuel peat cutting.)

It was therefore very important to the farmer and crofter that cutting peat (and indeed 'grass-turf') should be as simple and quick as possible, and the best way to ensure that was to burn off the overburden of vegetation. Since this was often also likely to improve grazing, burning on moors and peat bogs was a regular event: presumably this, along with direct grazing itself, was what kept the drier ones from instantly regenerating with birch and other trees, as so many are doing today. Because it was convenient, the burning itself often took place at unlawful seasons when the bog was driest, in late spring or summer, with attendant risks: a fact that perhaps accounts for much of the Scottish legislation limiting muirburn from the fifteenth century onwards.

For the farmer peat-cutter, convenience was more important than the law. This was how a cutter in County Antrim described his experiences around 1932:

> I had to get a winter's firing for myself, and having been used to watching my father when I was a young fellow, and seeing what he had done, I followed his footsteps. He always picked out a nice spot in the bog where he would start and cut. So I burned a nice wee bit first. That was to make the turf easier cut with the spade. Burning was against the law. I was even caught by the police myself one time and it cost me £3-10-0.

---

[33] A. Fenton, 'Paring and burning', in A. Fenton and A. Gailey (eds), *The Spade in Northern and Atlantic Europe* (Belfast, 1970), pp. 155–93.

[34] Darling, *West Highland Survey*, pp. 300–1.

He went on to explain how a friend who worked alongside was determined not to be caught by the police in the same way, and always lit the moss before he went to dinner, so that he could pretend it was an accident if the police came and he was not there:

> One day when he got back and the moss was burning he saw Mr Wilton [the policeman] watching and he ran as fast as he could and he got his shovel and he started beating out the fire and throwing water on it from the drain . . . Wilton got over and demanded what he was doing and he said: "Well you might ask. This was all right when I went for my dinner and look at it now – and even some of my turf has been burned."

He had the presence of mind to bluff Wilton that somebody else had lit the moss on him and Wilton just threw off his tunic and fell to and helped him to put it out and said nothing.[35] No doubt the village constabulary in Scotland had many equally trying experiences when people's treatment of the bog did not quite keep within the law.

---

[35] P. Smyth, *Osier Culture and Basket-Making: a Study of the Basket Making Craft in South West County Antrim* (Lurgan, 1991).

Fig. 7 The family pony with peat stack for a year, which it had helped to transport from the bog, Orkney, 1889. The huge resource of fuel and the ready availability of cheap horsepower are what kept the remote parts of Scotland and Ireland so heavily populated from the seventeenth to the nineteenth centuries. Photograph: Valentine Collection. Courtesy of St Andrews University Library: JV-11178.

CHAPTER 7

Energy Rich, Energy Poor: Scotland, Ireland and Iceland, 1600–1800*

'Three cold, miserable countries,' said Louis when he heard the title of this paper as it was delivered in a preliminary version, and indeed what characterises them are their high northern latitudes (Scotland and Ireland are on a parallel with Labrador, Iceland with Baffin Land), and their extreme oceanicity, which modifies the cold with rain-laden winds from a relatively warm sea. Windy, wet, overcast, cool, with short growing seasons but small temperature differences between winter and summer, our three countries were uncomfortable but easily habitable. They were, of course, also significantly different, Ireland having a small Mediterranean element in its flora (though whether original or introduced is disputed) and Iceland being so close to the Arctic Circle that its interior can be vegetatively classified as tundra. And they were markedly different within themselves, the north of Iceland being much colder than the south, and the east of Scotland and Ireland much drier than the west.

The sun was (usually indirectly) the ultimate source of all usable energy in all the countries of the earth in the early modern period, as it is the source of almost all of it today, except on Iceland where subterranean thermal sources have been utilised to such success in the twentieth century. In these climes, the direct warmth of the sun was never such as to obviate the need for domestic heating, as it was in the tropics, or to provide solar evaporation for a salt industry, as it did as far north as Vannes; nevertheless the sun pours the energy equivalent of 22,000 million tons of coal a year onto Britain, and even though only between one and four parts per thousand are harnessed by plants in photosynthesis, this has been enough to grow forests at these

---

* This formed a contribution to a volume in honour of Professor Louis Cullen, the distinguished and innovative economic and social historian who for many years occupied the chair of Modern Irish History at Trinity College, Dublin. It is reproduced from D. Dickson and C. O'Grada (eds), *Refiguring Ireland: Essays in Honour of L. M. Cullen* (Dublin, 2003), pp. 19–36. Thanks to Antony Parnell of The Lilliput Press for permission to reproduce.

latitudes.[1] The living plants are the source of energy calories in our food, the dead ones often the source of fossil fuels. Furthermore, the sun-driven energy of the wind and the rain is only too evident in our countries.

It is also important to note that energy in the past was not as interchangeable for man's use as it is now. Water falling over a dam could once do no more than turn a waterwheel: now it can generate electricity that can warm homes, power factories and run trains, or warm greenhouses for growing tomatoes. In particular there was in the past a firm line set, most obviously in industrial processes, between mechanical (kinetic) energy and thermal (fuel) energy: the latter could only begin to be transformed into the former when the steam engine was applied by Watt to rotary motion. In the early modern societies that we are considering the main forms of mechanical energy were human and animal muscle power, wind and water; the main forms of thermal energy were derived from peat, coal and wood: but wind could not warm you, and coal could not propel you. So to avoid energy shortages, there had to be sufficiency in several forms.

The other important consideration is the efficiency of energy utilisation. We need energy to work and food is our fuel, but if a man cannot raise more than 3,500 calories of food by a hard day's labour on the land, then he is reduced to absolute subsistence, since his energy expenditure and energy intake are no more than exactly in balance. On the other hand, if he sits at a tractor that uses fossil fuel he can raise vastly more food than he personally needs, partly by the expenditure of his own energy but mainly by that of energy released from the photosynthesis of earlier millennia. He can sell some of that surplus food to buy more energy of the latter kind in the form of diesel oil, and more of the food to buy consumer goods produced by energy. So with increased efficiency in energy utilisation, both his consumption of energy and his standard of living will rise.

It is the contention of this essay that pre-industrial Scotland and Ireland were energy rich, in that they had available to them extremely abundant supplies of usable energy, but that Iceland was energy poor. On the other hand, the energy wealth of Ireland and Scotland was often reduced by inescapable inefficiencies in utilisation. This meant that they were good places for many people to live, but not necessarily for many people to live well.

It is convenient to divide the subject into three parts: agricultural energy, domestic fuel energy and industrial energy. The potential of a country for agricultural energy – its ability to produce food – is not quite the same as its

---

[1] E. A. Wrigley, 'Meeting human energy needs: constraints, opportunities and effects', in P. Slack (ed.), *Environments and Historical Change: the Linacre Lectures 1998* (Oxford, 1999), p. 77.

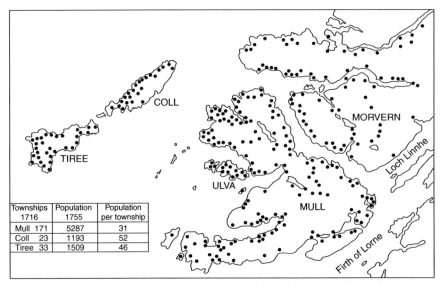

Townships 1716	Population 1755	Population per township
Mull 171	5287	31
Coll 23	1193	52
Tiree 33	1509	46

Map 1 Townships on Mull, Coll and Tiree. Adapted from N. Maclean-Bristol (ed.), *Inhabitants of the Inner Isles, Morvern and Ardnamurchan, 1716* (Scottish Record Society, Edinburgh, 1996) with permission.

fertility, though with sun and water naturally fertile land will obviously be ideal for food production, and this is as true of Leinster and Lothian as it is for the Netherlands and the Nile delta. But – and this is what is significant for our countries where so much of the ground is ranked of low fertility on soil maps, whether because it lies on steep slopes, has rock near the surface, or is leached, acidic and podsolised – relatively infertile ground can often also have high potential for agricultural energy production, though there are serious inefficiencies in its utilisation.

This can best be explained by illustration. Map 1 shows townships on the islands of Mull, Tiree and Coll in the Inner Hebrides, with some of the adjacent mainland. It refers to the situation in the first half of the eighteenth century: the townships, and their men of military age, were recorded at the disarming that followed the Jacobite rising in 1715, and the total population of the islands was estimated by the ministers for Webster's survey of 1755.[2] To those who know the islands now, what verges on the incredible is the density of settlement in what we see as the most inhospitable of environments – and all this before agricultural reorganisation and the coming of

[2] N. Maclean-Bristol (ed.), *Inhabitants of the Inner Isles, Morvern and Ardnamurchan, 1716* (Scottish Record Society, Edinburgh, 1996); J. G. Kyd (ed.), *Scottish Population Statistics including Webster's Analysis of Population, 1755* (Scottish History Society, Edinburgh, 1975). I am most grateful to Major Maclean-Bristol for permission to adapt his map.

the potato. Take, for example, the islands of Ulva and Gometra off the west coast of Mull, where there are today fewer than ten inhabitants and farming is being abandoned. In 1716, they had thirteen townships, sixty-eight men of military age and a total population probably between 300 and 400, all living entirely by agriculture.

Coll makes a particularly interesting study, as it was examined again in depth by Fraser Darling and his collaborators in the late 1940s.[3] The population had dropped from about 1,200 in 1755 to 210, having reached a maximum of 1,442 in 1841. Of its 18,300 acres, the total inbye extended only to about 1,900 acres, though in 1811 the arable and meadow acreage was estimated at 4,500 acres. What is significant in comparing the 1716 map with the 1940s is that the old settlements were precisely where the modern arable is, with the exception of about three destroyed by drifting sand; basically they are, and were, coastal and mainly on the west. Fraser Darling explains why:

The island of Coll is composed of Archaean gneiss . . . which does not of itself break down into good soil . . . the rock is in general covered with peat only . . . but approximately one-sixth of the island is covered by, or considerably affected by, shell sand blown in from the Atlantic. This sand with its high lime content neutralizes the acidity of the peat wherever it occurs and allows a flora characteristic of calcareous soils to develop.[4]

The population of Coll existed at its surprisingly high level because the energy of Atlantic storms scattered the skeletons of previous living creatures, which in turn had gained their energy from the sun, onto the thin peat, itself composed of fossil plants. There was even more to it than that, as the ground was further fertilised by kelp dragged up from the shore by people and animals: and kelp grows in seas warmed by the Gulf Stream and the direct light of the sun and also is flung on to the shores by the energy of Atlantic storms, though not carried so conveniently far as the sand. The climate of Coll was so mild that cattle did not need to be housed in winter: the grazing, though rough, was extensive. Sunshine levels for Coll and Tiree have always been above the average for the Hebrides as a whole.

Boswell came with Dr Johnson to Coll in 1773 and described the 'great number of horses', the black cattle 'reckoned remarkably good', the

---

[3] F. F. Darling (ed.), *West Highland Survey: an Essay in Human Ecology* (Oxford, 1955), pp. 377–400.
[4] Ibid. p. 379.

numerous sheep 'which they eat mostly themselves', the goats, the lochs with trout and eels, the wildfowl, the pigeons, plovers and starlings, 'of which I shot some, and found them pretty good eating', their home-grown dyestuffs, their imported tanbark and bonnets. It is a picture of an island that bought and sold little externally, though some horses and cattle were exported – a community that lived by farming and a little hand-line fishing; it was poor, but in no way did it seem to Boswell and Johnson as horrifyingly wretched.[5]

Coll was typical of hundreds of communities down the west coasts of Scotland and Ireland. Each was different, and had its own little advantages: on Gometra and Ulva there was no blown shell sand, but there was a basaltic soil; on Lewis there was boulder clay beneath the peat, if you could reach it. Down much of western Ireland there was limestone, and everywhere there was the power of the Atlantic. In 1814, the Commissioners for Irish Bogs described the distribution of population in Connemara: nine-tenths were along the seashore. Shell and coral sand was raised by dredging or by beaching a boat at low tide (the fury of the wind evidently did not drive it so far inland as in the Hebrides). Red seaweed was abundant, both cast ashore and cut from boats:

Two or three boat loads of about six tons each are usually applied as manure over an acre of potato ground . . . the value of the sea manure is abundantly shown by the numerous patches of cultivated ground which occupy the shore from Galway westward, and where the soil must originally have been of the most uninviting description, being nothing but bog and rock.[6]

Such localities had devised their own cereals, the hardy 'bear' or primitive barley of the Hebrides, and in more favoured spots, oats, but when the potato came, it replaced the cultivation of grains as the main way of producing energy from crops: the calorific yield of an acre under potatoes was three or four times that of an acre under oats. The potential for the production of agricultural energy suddenly rose as well, though to raise the crop on thin soils on rocky shores demanded labour-intensive spade husbandry, so that the ratio of human energy input to food energy output was not necessarily any more favourable. To quote Kevin Whelan, speaking of potato husbandry in rural Ireland:

[5] J. Boswell, *The Life of Samuel Johnson to which is added the Journal of a Tour to the Hebrides* (London, 1897), pp. 617–24.

[6] *Fourth Report of the Commissioners appointed to enquire into the nature and extent of the several bogs in Ireland* (London, 1814), p. 10.

There was five times as much labour expended in digging rather than ploughing an equivalent area, and twenty times as much as was expended in stock rearing. The considerable energy costs necessitated by spade cultivation on marginal land were absorbed by population growth. In a sense then, as in Highland Scotland, increased population became its own resource.[7]

What the energy costs amounted to in terms of time spent in the field can be shown in Shetland, where, using a cas-chrom, it took one person from Christmas until late April or mid-May to till enough ground to keep a family of seven or eight in potatoes and meal for a year.[8]

But with more agricultural energy, contemporaries looked forward to more population. The one enabled the other. James MacDonald in the *General View of the Agriculture of the Hebrides* in 1811 announced that 'the produce of potatoes ... promises, by tolerable management, to triple the number of inhabitants every 75 years without burdening the country'.[9] We all know (but only from hindsight) how little such a sanguine view was justified. Nevertheless, perhaps historians have not sufficiently dwelt on how it was possible for such large populations, both before and after the arrival of the potato, to thrive in what seem to us such difficult places as the west of Ireland and Scotland.

Three additional things, that are not usually stressed, might help to explain it. The first is the widespread distribution of white clover as a naturally occurring leguminous herb throughout Ireland and the west of Scotland. This has been referred to in the past, but it is worth re-emphasis given the startling improvements in fertility attributed to sown clover in less favoured areas like eastern Scotland and Denmark in the eighteenth century.[10] The second is the fact that in order to raise food energy from crops that grow and need to be harvested between April and October, you do not actually have to spend a full day's energy every day or all the year round. A full 3,500 calories are needed for a fit man ploughing, carrying seaware, digging and leading peats (of which more later) and harvesting the crop, but for periods energy expenditure could fall to a much lower level without

---

[7] K. Whelan, 'The modern landscape: from plantation to present', in F. H. A. Aälen, K. Whelan and M. Stout (eds), *Atlas of the Irish Rural Landscape* (Cork, 1997), p. 88.

[8] A. Fenton, *Scottish Country Life* (Edinburgh, 1976), pp. 43–5.

[9] Quoted in M. W. Flinn (ed.), *Scottish Population History from the Seventeenth Century to the 1930s* (Edinburgh, 1977), p. 427.

[10] See also Chapter 8; T. Kjærgaard, *The Danish Revolution, 1500–1800: an Ecohistorical Interpretation* (Cambridge, 1994), pp. 84–7; C. P. H. Chorley, 'The agricultural revolution in Northern Europe, 1750–1880: nitrogen, legumes and crop productivity', *Economic History Review*, 34 (1981), pp. 71–93.

involving starvation. It could fall to as low as 1,500 calories, which is what is needed in a daily diet to meet basic metabolic needs. At that point a person becomes lethargic, and as a strategy for saving energy it is what animals do all the time as a way of husbanding their expenditure of calories. There are plenty of accounts, usually by the hyper-well-fed visitor from outside, of Highlanders and Irishmen being indolent, but none of them being fat: arguably they are pursuing an energy-conservation policy that was not in itself seen as a hardship. Of course there was also a tradition, very obvious in the nineteenth century but at least in Highland Scotland traceable to the seventeenth century, of migration away to the Lowlands and to England to work long and hard as harvest hands and navvies. But that makes it easier still for the rest of the family to exist in County Clare and Skye, if the hungry ones remove themselves from the scene and get their energy needs met for part of the year at someone else's expense.

The third point is that in these regions, horses, at least at some times and in some places, were practically a free good. When Dr Johnson landed on Coll, his hosts, in Boswell's words:

> ran to some little horses, called here *Shelties*, that were running wild on a heath, and catched one of them. We had a saddle with us, which was clapped upon it, and a straw-halter was put on its head. Dr. Johnson was then mounted, and Joseph very slowly and gravely led the horse.[11]

In 1811, it was said that there were 500 horses on Coll (compared to 1,200 cattle).[12]

So it was over much of upland and western Scotland and western Ireland, and also over much of Iceland. The horses (some would now call them ponies, though I do not think that was contemporary usage) ranged on open, rough ground at minimal opportunity cost in respect to the available grazing, and never had to be housed indoors in winter like other domestic beasts, even in Iceland. When summer grass died away, they ate heath and gorse, browsed on trees and shrubs or dug for roots, and in Shetland ate seaweed. In Iceland, 'when the usually scanty supply of hay comes to an end, they readily take to eating cods-heads specially reserved for them during the fishing season.'[13] They were reckoned to be part of the natural endowment of the country: when Timothy Pont first surveyed the north-west Highlands in the 1590s,

[11] Boswell, *Samuel Johnson*, p. 617.
[12] Darling, *West Highland Survey*, p. 384.
[13] J. C. Ewart, 'The ponies of Connemara', in W. P. Coyne (ed.), *Ireland: Industrial and Agricultural* (Dublin, 1902), p. 348.

he commented in his notes on the rough maps about mountains, woods, salmon, pearls and about horses – 'item in Assyn in Laid Markill, wher ar guid stoods', and 'ther ar many lusty stuids in Brachat and Stranaver'.[14]

Wrigley has dwelt on the energy benefits of draught animals in an agricultural society: 'a horse is on average about ten times stronger than a man . . . a horse-rich community was a mechanical energy-rich community, which would consequently probably also be rich in the conventional sense'.[15] In the three societies we are considering, these benefits were still real, though severely limited by certain factors. Firstly, to get the best out of a horse you have to give it supplementary feeding of hay or oats for the most strenuous tasks, and it is doubtful if our three societies had any to spare: so they would be low-efficiency animals. Secondly, the horses that could survive in the wild, or semi-wild, were rather small: a Sheltie was very much less than ten times the strength of a man. Thirdly, where land was so thin and rocky that much of it could only be cultivated by spade or cas-chrom, a horse is of limited use for ploughing or harrowing, despite locally adapted implements designed to jump over rocks in shallow soils.[16] But not all the Highlands or Ireland were like this, and in Kintyre, for example, horse-ploughing was in use at a time when oxen were still the main plough animals in much of the Lowlands.

The main use of horses in all three countries was for carriage. The Icelandic horses were described as carrying loads of fish of 300–400lbs each from the fishing stances to the export harbours, in trains of twenty to thirty laden beasts and a few others unburdened to be used if some collapsed.[17] The Sheltie in its native islands was principally used to bring seaware from the shore to the land, and its use for this purpose must have substantially extended the area that could be thus fertilised. They were also used for carrying peat, and miscellaneous heavy loads such as Dr Johnson. At the beginning of the twentieth century, J. C. Ewart explained the varied duties of the Connemara pony according to the time of year:

At one time they may be seen climbing steep hillsides heavily laden with seaweed, seed corn or potatoes; at another they convey the produce to market. Sometimes it is a load of turf, oats or barley; at other times creels crowded with a lively family of young pigs. During summer and autumn the ponies are often seen trudging unsteadily along, all but buried in a

---

[14] J. C. Stone, *The Pont Manuscript Maps of Scotland: Sixteenth Century Origins of a Blaeu Atlas* (Tring, 1989), p. 31.

[15] Wrigley, 'Meeting human energy needs', p. 80.

[16] Fenton, *Scottish Country Life*, p. 36.

[17] Uno von Troil, *Letters on Iceland* (London, 1780), p. 57.

huge pile of hay or oats . . . Returning from market each pony generally carries two men, one in front and the other on the pillion behind.[18]

While the abundance of small horses was not enough to tip the balance and make our areas 'rich in the conventional sense', they must have done something to redress other disadvantages of natural endowment. 'Without a pony the peasant farmer in the west of Galway is all but helpless,' Ewart summed it up at the start of the twentieth century.[19]

So the argument is that Scotland and Ireland were rich in agricultural energy, or rather in the potential to realise it, even in their least outwardly fertile and most oceanic districts. Thanks to the power of the sun and its minions wind and rain, they had the potential to grow crops and fodder enough to support locally dense populations, they had much free fertiliser from the sea, plus clover, and they had pony power. Iceland, by contrast, was energy poor, despite its wealth of horses and its ability to support sheep and cattle. Due to climatic deterioration, the sun did not provide warmth enough to grow grain crops to any great extent after the fourteenth century. Potatoes were introduced at the end of the eighteenth century but never became more than a garden crop, possibly because of the intensity of winter frosts.

There was, however, a critical energy transfer from the fertile sea, the inshore waters accessible by open rowing boat, in terms of calories from protein and fat. Dried fish formed a bread substitute in Iceland between 1200 and 1900, eaten with sour butter. The oil of fish, seals, basking sharks and whales was drunk separately. Of course, there was also fishing in Ireland and Scotland, and the economy and diet of Shetland has some parallel to that of Iceland, but in Iceland it was very much more critical for survival in far more marginal circumstances.[20]

Living by livestock, supplemented by what could be gleaned from the sea by fishing and whaling, the population of Iceland stagnated for centuries. It failed to exceed 50,000 (with periodic crashes due to disease and famine well below that – the lowest recorded was 38,000 in 1780) and did not move above that ceiling at all in the early modern period, or until after 1820: it is now at ten times the eighteenth-century level.[21] At the same time, the rural populations of Scotland and especially Ireland, particularly in the most oceanic areas, grew rapidly. The population of Ireland grew perhaps sixfold between 1600 and 1840, at which point population density per square

---

[18] Ewart, 'Ponies of Connemara', p. 349.
[19] Ibid. p. 349.
[20] I am indebted to Ole Lindquist for these points about the energy transfer from the sea.
[21] E. Laxness, *Islandssaga* (Reykjavík, 1995), vol. 1, pp. 12–13, 116; vol. 2, 147–9. I am obliged to O. Lindquist for these and other citations.

kilometre was probably ten times that of Iceland. 'There is scarcely any country so little favoured by nature,' said the Swede, Uno von Troil, who accompanied Sir Joseph Banks on his expedition to Iceland in 1772.[22] If the sun withholds its warmth, a society lacks the energy to prosper.

To turn next to domestic fuel energy, let us begin with Iceland. Here was a country in extraordinary danger in the eighteenth century of running out of the means to keep warm. In the words of an eleventh-century Icelandic historian, when the island was discovered in the latter half of the ninth century, it was 'covered by woodland from the mountainsides to the sea-shore', and modern studies have postulated that two-thirds of the country was vegetated and half that vegetation was woodland. By 1800, woodland covered only 4 per cent of the land surface and was in very poor condition, having been reduced by centuries of overgrazing and unsustainable use.[23] Given the small extent and often inaccessible location of the remnant woods, this meant that they were not readily available for domestic fuel. It is really difficult to understand why Icelanders did not do more to conserve sufficient of their birchwoods to supply them with fuel. Admittedly, birch does not coppice well and will not regenerate under its own shade, so the usual ways of sustainably managing woodland seen in the British Isles or southern Europe were less appropriate. But it regenerates readily in enclosures next to existing birchwoods, and one could expect the emergence of wood-lot rotation serving each village. This would have implied the sacrifice of a certain amount of pasture, and possibly the rigidities of a tenurial system where rent was paid in fixed amounts of wool may help to explain an inability to sacrifice any grazing at all. Without wood, Icelanders had recourse to limited quantities of a very inferior peat and a little lignite, the latter available only to a very small number of farms. For many on the coast even peat was not available, and the inhabitants there burned other vegetation such as fern, juniper and crowberry, supplemented by cattle bones, fish moistened with train oil, dried animal dung and seaweed. Wrack was cut off the rocks with sickles and on certain farms as many as 400–500 horse-loads a year were transported home from the beach for use as fuel.[24]

The situation was the more ironic because Iceland sat on an immense store of thermal energy from volcanoes and hot springs, which, if it could have been tapped, would have transformed them from energy paupers to fuel-energy millionaires. If you go to Iceland today, you will encounter very warm houses

[22] Troil, *Iceland*, p. 25.
[23] S. Blöndal, *Vegetation and Prosperity: the Struggle to Reclaim Iceland's Devastated Land* (Forestry Fund, Reykjavík, 1994), pp. 2–4.
[24] L. Kristjánsson, *Íslenzkir Sjávardhættir* (Reykjavík, 1980), vol. 1, pp. 136–40 [English summary, p. 444]; Troil, *Iceland*, pp. 42–5.

looking much like those elsewhere in Scandinavia and the hottest showers you are ever likely to find anywhere. However, apart from some use of hot springs for laundries and public baths in Reykjavik in the eighteenth and nineteenth centuries, no serious use of this resource was made before the 1920s.

The extreme shortage of domestic fuel led Icelanders before the twentieth century to construct the most energy-efficient houses in Europe, but not the most comfortable. They were built above ground but essentially of lava, earth and turf so that sometimes they had an entirely subterranean appearance. Driftwood was used for the doors and internal wainscoting, possibly for one gable end. Windows, if any, were minimal and 'composed of the thin membranes of certain animals'. Stock shared the houses, even of the most prosperous farmers, in an even closer intimacy than was the case in the homesteads of poor cotters in Scotland and Ireland. Troil commented that the houses had no chimneys, as Icelanders never lit a fire 'except to dress their victuals, when they only lay the turf on the ground', the smoke escaping from a square hole in the roof.[25]

A colleague in Iceland sends the following account:

Icelandic peat is poor and was hardly used, firewood from scrub precious (driftwood does not burn), so as of c.1100 the houses here became earthen houses and were essentially unheated until 1870-80. The earthen houses ([in] 1940, one third of roughly 5,000 farms still had such living quarters) were connected with corridors, the floor of which rose a little towards the inner rooms to direct the little warmth in their direction. The cow and sheep sheds were at the front, then came the kitchen and at the very back the common living room of master and all servants (women at the outer wall!), children with the women and at the feet of the elderly (to keep them warm). In the southeast, the living room was placed over the cow and sheep sheds.

The compression of the living space resulted in people having to eat all meals from a wooden bowl with a lid while sitting on their bed, because since the high Middle Ages there was no space for tables. In the Icelandic annals, marvellous sources until 1800, we often read the remark by the vicar: 'People died for lack of fat'. The meaning is that dried fish and even meat was available (protein), but when the people could not keep up their individual body heat, things went really wrong ... As late as the 1970s I encountered the reminiscences of this emphasis on fat in the daily food.[26]

---

[25] Troil, *Iceland*, pp. 27-8, 99-101.
[26] O. Lindquist, *pers. comm.*

In Ireland and Scotland, on the other hand, domestic fuel was abundant, though with some patches of local scarcity. The Central Belt of Scotland with the main towns of Edinburgh and Glasgow had plentiful coal mined within easy reach and it could be shipped coastwise to Aberdeen or Dumfries relatively cheaply once the internal coal duties were lifted in Scotland in 1793. One of the advantages of coal was that it had a remarkably good energy input/output ratio – a miner could produce in a day a heat value at least 500 times greater than the heat value of the food he consumed.[27] Eastern Ireland also imported coal from Britain fairly cheaply and on a large scale. Some 40,000 tons were exported from Britain to Ireland in 1700, more than doubling to 98,000 tons in 1750; 399,000 tons in 1800; and to 717,000 tons in 1825–8.[28] Export dues on coal to Ireland before the 1770s were actually lower than on coastwise trade in Britain. Some of this coal was destined for industry like sugar-refining, brick-making, brewing and glass-making, but by far the larger part must have been for domestic use. Mercantilist minds worried about Ireland's dependency on external fuel, but it was Dublin that had Whitehaven and its rivals by the forelock in commercial terms rather than the other way round, and intense competition between coal-producing ports in England, Scotland and Wales, some without obvious alternative outlets than Ireland, kept the price of warming a Dublin house comparatively low. In the early 1840s, a ton of coal in the main Irish ports cost 11s. or 12s., in Edinburgh about 8s. and in Glasgow 5s.; in Aberdeen it cost about 20s. and in Dumfries about 14s.[29]

Coal was also extraordinarily convenient and compact compared to alternatives. Average bituminous coal yielded per ton well over twice the energy of air-dried wood, and at least twice that of peat. In order to replace 400,000 tons of coal, Ireland would have needed to keep over 200,000 hectares of woodland on coppice rotation, or to move annually to points of urban consumption more than an additional 800,000 tons of peat.[30] As it was, Dublin did import substantial quantities of peat, some 30,000 tons a year by the Grand Canal from the Bog of Allen alone in the early nineteenth century, and a good deal more by cart. Overall, Ireland was much more dependent on peat

---

[27] Wrigley, 'Meeting human energy needs', p. 87.

[28] M. W. Flinn, *The History of the British Coal Industry* (Oxford, 1984), vol. 2, p. 222.

[29] J. Mokyr, *Why Ireland Starved: a Quantitative and Analytical History of the Irish Economy, 1800–1850* (London, 1983), p. 154; I. Levitt and C. Smout, *The State of the Scottish Working-class in 1843* (Edinburgh, 1979), p. 62.

[30] See T. Johnson, 'The Irish peat question', *Economic Proceedings of the Royal Dublin Society*, 1 (1899), p. 26; Wrigley, 'Meeting human energy needs', pp. 86–7. The assumption here is that each hectare of woodland annually produces by photosynthesis the equivalent of 2.5 tons of coal (50 tons in a twenty-year rotation) but that due to wastage of twigs and leaves only 75 per cent could be harvested.

than on coal. It is said that during the eighteenth and nineteenth centuries at least 5 million tons a year was dug out and, even as late as the 1920s, 6 million tons of peat were being taken by hand for domestic use, as opposed to about 2.25 million tons of coal imported for both industrial and domestic uses.[31]

Nevertheless, one consequence of the abundance and simplicity of coal supply was that the new eighteenth- and nineteenth-century housing of the middle ranks in Dublin and Edinburgh, as of Glasgow and Belfast, was as energy-inefficient, and as gracious, as any in Europe: big windows on all sides, facing up to the prevailing winds or to the sea view, uninsulated roofs of slate, open hearths with coal-burning grates that mainly warmed the chimney, they were monuments to conspicuous fuel consumption. The only difference that can be detected is that eighteenth-century Dublin houses sometimes had the upper walls insulated with sphagnum moss, which possibly reflects the slightly higher costs of fuel in Ireland.

Most of the rest of Ireland and Scotland which did not have ready access to coal had, as we have seen, abundant peat, often deep and of excellent quality, commonly regarded as inexhaustible and free but for the labour of digging it. Such labour was quite considerable. Dug out wet, about eight tons of raw peat were needed to yield one ton of air-dried peat; in 1921 about three tons of air-dried peat were used in Ireland per head per year, by the peat-using part of the population. As peat had half (or less) the calorific efficiency of coal, over sixteen times the human energy had to be exerted at the face to extract a given number of fuel calories from peat as from coal. It was reckoned that an expert cutter could dig enough peat to provide three tons equivalent air-dried in a day, but few cotters or crofters would have worked at that pace in practice.[32] Additional labour was needed to transport it to the drying stance, and then more to carry it dried to the home or to market. It was often seen as a pleasant communal effort, but it was effort nevertheless.

An alternative calculation takes account of the number of 'loads' that were required to keep a fire for a household for a year. In County Tyrone, in 1802, the quantity was reckoned at sixty 'kishes', a kish being described as being equal to a cubic yard but the kish basket was heaped as high as could be conveniently carried.[33] In Moray and Nairn and in the Outer Hebrides, fifty to sixty loads were required annually, and the figure of forty to fifty

---

[31] *Report on Peat* (Commission of Inquiry into the Resources and Industries of Ireland, Dublin, 1921), p. 8; J. Feehan and G. O'Donovan, *The Bogs of Ireland: an Introduction to the Natural, Cultural and Industrial Heritage of Irish Peatlands* (Dublin, 1996), pp. 2, 8; O'Gráda, *Ireland, a New Economic History 1780–1939* (Oxford, 1994) puts the amount of peat dug in Ireland in the nineteenth century at 5–6 million tons annually.

[32] *Report on Peat*, pp. 8, 20, 26, 98.

[33] J. McEvoy, *Statistical Survey of the County of Tyrone* (Dublin, 1802), p. 151.

loads is mentioned for County Donegal.[34] In North Uist, the cutting, drying and carting of this quantity was equivalent to a full month of a man's time. It sounds a lot, but I would guess that in many homes today the fuel bills are equivalent to a month's salary – and compare this to the 300 to 400 horse-loads needed for the seaweed fuel in Iceland.

There was, however, much variation, and the estimates quoted so far were not the only ones. One interesting Scottish calculation of around 1940 reckoned the labour cost of providing a year's fuel 'for a household' in Strathpeffer, Easter Ross – 'in no way exceptional and [which] may serve as a general guide'. In this case, the peat bank was three and a half miles from the house and a mile from the nearest road. In all, 708 man-hours and 336 horse-hours were needed to obtain sixteen tons of peat, reckoned in this case to be equal to only four tons of coal (half the usual calculation in Ireland). This represented fifteen work days (at twelve hours per day) for four people, and the work of two horses for fourteen days: 'one man might do the work of extraction in something like three weeks; the rest is carting'.[35] Distance to the peat was therefore a very important factor, as well as the quality of the peat.

It is therefore unsurprising that the homes even of the tenantry were not built to be profligate with fuel in peat-burning areas. Sharing them with animals was an energy-saving choice for small farmers and cotters just as it was for almost everyone in Iceland, but at least they had four walls clear above ground built of turf or stone, with wooden crucks and panels of withy, and a constantly burning hearth often warming well a considerable interior space.

It is a mistake to suppose that the Irish and the Scots only turned to using peat when local forests were cut down and wood became hard to get. Peat was in every way a superior fuel to wood. A year's supply was cut with less labour. It needed only one tool that did not need sharpening or replacing. Wood was difficult to process and hard to transport, and not clearly of calorific advantage, ton for ton. It took even longer to dry. Peat was safer, spark-free and flexible to use in the home.[36] Nations like the Swedes and the French who had nothing to burn but their forests were, in respect to domestic fuel, energy poor in comparison to the Irish and the Scots.

One can no doubt overdo the picture of an energy-rich Scotland and Ireland in terms of domestic fuel, for there were often patches of real poverty

---

[34] Fenton, *Scottish Country Life*, p. 197; Congested District Boards 'Base-line reports' (printed but unpublished, 1891) [kept in Trinity College Dublin].

[35] A. Collier, *The Crofting Problem* (Cambridge, 1935), p. 187.

[36] A. T. Lucas, 'Notes on the history of turf as fuel in Ireland to 1700 AD', *Ulster Folklife*, 15–16 (1969–70), pp. 172–4.

where, for one reason or another, peat or coal had become unavailable. The inhabitants of Tiree imported their peat from Mull, and those of Iona had their own peat bank across the strait on Mull, which they cut for themselves. There were instances in Orkney where, like Iceland, seaweed and animal dung were used for fuel. In many places in the *Statistical Accounts* of the 1790s, there were instances of difficulty on the mainland – even of local depopulation, where peat banks close to the settlement had been worked out and people might have to go for a mile or much longer to get their fuel. Turf, heather or broom in such places might be utilised as inferior substitutes for peat. Wood was seldom a possibility as, by 1800, the proportion of land under wood in Scotland and Ireland was small, as it was in Iceland. When – as in Scotland in the eighteenth century – landowners and farmers enclosed or appropriated open ground that had been used by the poor to get their fuel, the consequence for the latter could be serious. Fenton cites the case of Eccles in Berwickshire in 1794, when the distress of the poor was 'unspeakably great' and of Tongland in Kirkcudbrightshire, where there was 'real distress for fuel, which many poor and shivering wretches suffer'.[37]

Similarly, in Ireland it was reported in 1891 of the parish of Clonmay in Donegal that peat had to be fetched from long distances – 'the first famine to occur in this district will be a fuel famine', and at Glencolumbkille in the same county that the peat was so distant from some town lands that people spent up to three hours a day to 'creel down the turf on their backs.'[38] Estyn Evans described the consequences in Ireland of a wet summer when the peats would not dry, followed by a hard winter: then 'tragedy may follow', the countryside stripped of what little timber it could provide and everything from furniture to fishbones and cow dung used to boil the pot. In bog-less districts in the seventeenth and eighteenth centuries dung was regularly used as fuel.[39]

But such localities remained exceptional in Scotland and Ireland. The converse was of course true in Iceland, but to a lesser degree, and not only there. For example, Thorkild Kjærgaard has argued that eighteenth-century Denmark suffered from a crisis of cold brought about by a growing shortage of fuel-wood and the absence of much alternative except poor-quality peat and turf. To quote him:

The Danish population has never in historic times suffered as much from the cold as it did during the last five or six decades of the eighteenth

---

[37] Fenton, *Scottish Country Life*, p. 194.
[38] Congested Districts 'Base-line Reports'.
[39] E. E. Evans, *Irish Folk Ways* (London, 1957), p. 192.

century and the first three or four decades of the nineteenth – even though people kept their hats, their coats, and their boots on indoors and sometimes even in bed. At no other time in Denmark have there been so many indoor walls glistening with frost, so many draughty floors, so many dripping noses and so many people suffering from the cold and rheumatism as there were during those endless years.[40]

Were not Scotland and Ireland lands of luxury compared to that?

Given the plenitude of agricultural energy and domestic fuel energy, Scotland and Ireland but not Iceland might, to use an eighteenth-century phrase about Carolina, be candidates to be 'the best poor man's country in the world'. But to get beyond that, to get growth, development and be a rich man's country as we are all today, industrial energy was more important.

Industrial energy, in the centuries and countries that we are considering, came from kinetic energy (water, wind) and fuel energy (wood, peat and coal). The two were scarcely interchangeable until after 1800 when Watt's patent expired. Again, to take Iceland first, it had abundant water and wind, but very little use for it. Grinding grain into meal or flour was the usual *raison d'être* of a watermill or a windmill in an agricultural society, but there was no grain cultivation after the Middle Ages, and though watermills are mentioned in the sagas nothing is found thereafter until, apparently around 1900, one was established in a creamery. Windmills are similarly absent until the nineteenth century, when two or three were built. Wood, as we have seen, was in extremely short supply, but charcoal was nevertheless needed for hammering the hay-making sickles made out of soft iron, sharp again: 'had it not been for the Scottish steel scythe (imported in the nineteenth century)', writes my Iceland colleague, 'the very last patches of Icelandic birch scrub would have been gone'.[41] As for coal, there was none, and peat as a substitute was so scarce it had to be kept for such domestic heating as Icelanders could contrive. So Iceland was simultaneously oversupplied with kinetic energy that it could not use and a pauper for fuel energy that it needed. The contrast with the position today is again astonishing, where there is such a surplus of thermal energy and hydropower that engineers in the 1990s dreamed of exporting it by submarine cable to the European Union – a project now abandoned in favour of utilising it for ambitious aluminium smelters in Iceland.

Scotland and Ireland had as much rain and wind as Iceland, but more use for kinetic energy harnessed by mills. Until the eighteenth century,

---

[40] T. Kjærgaard, *Danish Revolution 1500–1800*, p. 97.
[41] Lindquist, *pers. comm.*

watermill technology in both countries was largely confined to grinding meal and flour, and they were a common adjunct to semi-subsistence agriculture. Horizontal watermills, operating on a very small scale, often for a single farm, were introduced into Ireland perhaps as early as the third century AD. In both countries, these were restricted to the wettest places with small hilly streams, and in some localities were extremely numerous. Sir Walter Scott estimated that 500 were still working in Shetland at the time of his visit, and certainly the six square miles of Papa Stour even now contain twenty-four mills and mill sites. Elsewhere the more sophisticated and larger mills operating with vertical wheels were used, often enforced on a group of farms as an adjunct to baronial thirlage, as in Orkney or Aberdeenshire. Many thousands of medieval and early modern grain-processing watermills of both kinds are therefore found the length and breadth of Scotland and Ireland.[42]

On the other hand, there were remarkably few windmills in either country, even in the eighteenth century. By 1840 in Ireland, there were about sixty in East Down and thirty in South Wexford, another eighty in the drier east-coast corn-growing areas and about forty elsewhere in the country. Compare that with 180 watermills but only one windmill recorded in County Kilkenny in the nineteenth century.[43] In Scotland, there was a scatter of windmills in the drier eastern counties from Fife and Berwick to Cromarty and Orkney but virtually none elsewhere, and some of those were for specialised purposes like pumping water or grinding lead ore. In truth, water power in a wet country was far preferable to wind power in a windy one, because the wind was too fierce and gusty. Nor, for all that impressive superstructure, does a windmill deliver much power: Smeaton set it as the equivalent to two-horsepower. Others have disputed this, but it has been asserted that even in the Netherlands, in the Dutch golden age, the thousands of windmills in operation provided less than 4 per cent of the energy gained from peat.[44]

Quite early on, watermills had been adapted for fulling cloth in Scotland: there were 300 known waulk-mills in the country between 1550 and 1730,[45] and Scotland was one of the sources from which water-powered fulling was introduced to Ireland in the seventeenth century. However, the most significant period of water power in industry fell between 1680 and some point

[42] Fenton, *Scottish Country Life*, pp. 102, 105, 107.

[43] Aålen *et al.*, *Atlas of the Irish Rural Landscape*, p. 226.

[44] R. W. Unger, 'Energy sources for the Dutch Golden Age: peat, wind and coal', *Research in Economic History*, 9 (1984), pp. 228–9. For a critique, see K. Davids, 'Innovations in windmill technology in Europe', in S. Cavaciocchi, *Economica e energie (secc. xiii–xviii)* (Prato: Instituto Internationale di Storia Economica 'F Datini', 2002), vol. 24, pp. 271–92.

[45] J. Shaw, *Water Power in Scotland, 1550–1870* (Edinburgh, 1984), p. 46.

in the first half of the nineteenth century. During that time, watermills and waterwheel technology came to be applied to an ever-increasing number of industrial processes in Scotland and Ireland, beginning with paper-making and culminating in textile spinning, especially cotton-spinning, but encompassing a whole range of other processes, such as washing, rubbing and beetling cloth in the bleachfields, scutching lint, saw-milling, brewing and distilling, the drainage of mines, and certain processes in the manufacture of iron, gunpowder and lime.

The great strides made by millwrights in water-power technology in the eighteenth century rendered extremely valuable, places that combined high and steady rainfall on a moderate slope with reasonable access to markets. These were the premium industrial sites in Europe, and Ireland and Scotland had many of them, ranging from the Lagan and Bann valleys in Ireland to New Lanark, Deanston and Stanley on the Clyde, Forth and Tay in Scotland. Nor did the effective transmission of steam to kinetic energy after 1800 completely alter the situation, since, as Bruce Lenman notes, just as the sailing ship reached its height of functional perfection in competition with the steamer, so the waterwheel was much improved after the initial spread of the steam engine.[46] In 1838, horsepower generated by steam in cotton and flax factories in Scotland exceeded that generated by water by two to one, but in woollen factories the reverse was true, and steam did not exceed water there until after 1860.[47] The nineteenth century also saw the largest complex of watermills in Ireland – often in what Fred Hamond attractively terms 'a seam of early industrialisation', such as that running along the river Dodder and its tributaries from the southern edge of Dublin to the fringe of the Wicklow mountains in 1837 – manufacturing cloth, paper, flour, pins, starch and so forth.[48]

The energy of water diffused industrial activity, and ensured there was often plenty going on in the countryside as well as in the town. That is perhaps one reason for the extraordinary consensus of optimism that pervaded Scotland and Ireland when their elites considered economic growth in the later eighteenth century. The landed classes, and the intellectuals whom they patronised, saw it as much less of a threat to their position than if it had been predominantly urban – as it became when coal replaced water. Although by the end of the eighteenth century more coal was still consumed in Scotland and Ireland by domestic than by industrial users, the latter had clearly been growing in importance. Its relatively cheap and plentiful supply

---

[46] Lenman, *An Economic History of Modern Scotland, 1660–1976* (London, 1977), pp. 115–16.
[47] Shaw, *Water Power*, pp. 524, 528.
[48] In Aålen *et al.*, *Atlas*, p. 229.

in the large towns of both countries plainly added to their attractiveness as industrial locations.

Coal supply has already been discussed, so it is appropriate to conclude by discussing that other important industrial fuel, charcoal derived from wood, which was long essential to one important industry, iron smelting, and where, for once, Ireland and Scotland have contrasting histories. There is a conundrum here. Why in Ireland was the use of woodland for making charcoal to fuel ironworks followed by a decline in woodland cover, while in Scotland such use was followed by more careful conservation and management of the resource? Wood, after all, was about the only renewable fuel known in early modern society. It was also the only fuel that could be used for the smelting of iron until well into the eighteenth century (and for some grades of iron until well into the nineteenth century), and adequate supplies of iron were essential to most activities from husbandry to warfare. Woods would seem worth keeping.

It is as well to outline the facts in both cases, as they have been widely misunderstood. Ireland at the time of the Elizabethan and Jacobean onslaughts and settlement was perhaps not largely covered with primeval forests of oak, but there was certainly a great deal more semi-natural wood than remained subsequently. McCracken's estimate of about 12 per cent of the land surface under wood may or may not be too generous, but even one half of that would still represent three times as much wood as existed in 1800.[49] The decline was associated with the manufacture of barrel staves and with shipbuilding as well as with iron smelting, but the latter and, to some extent, glass-making, are held in the literature to bear heavy responsibility for the exhaustion of the woods. To this Oliver Rackham adds a partially dissenting note, observing that where extensive seventeenth-century woodland did survive into the nineteenth century (as in Counties Kerry, Waterford and Wicklow) there is usually a history of ironworking, but he, too, would acknowledge that little of it survived thereafter and that other centres of iron manufacture in Ulster or County Leitrim were disafforested earlier.[50]

In Scotland, the charcoal ironmasters operated mainly along the west Highland coast, which was as well-wooded with oak as any part of Ireland. Interestingly this exploitation was largely pioneered by Irishmen, although they were not very successful,[51] and it was restarted with more success by

[49] E. McCracken, *The Irish Woods since Tudor Times: Distribution and Exploitation* (Newton Abbot, 1971); J. R. Pilcher and S. Mac an tSaoir (eds), *Woods, Trees and Forests in Ireland* (Dublin, 1995).

[50] In Pilcher and Mac an tSaoir, *Woods, Trees*, p. 3.

[51] T. C. Smout, A. R. MacDonald and F. Watson, *A History of the Native Woodlands of Scotland, 1500–1920* (Edinburgh, 2005), esp. ch. 13.

firms from Cumbria around 1750, some of which persisted until well into the nineteenth century, while the woods that they used survived and prospered. The Scottish case is the less unexpected. When the main English companies settled at Bonawe and Loch Fyne, they secured woods right down the coast and provided advice to landowners on their management. The result on the estate of the Duke of Argyll, for example, was a revolution whereby the woods changed from being used for grazing, winter shelter and as a peasant timber resource, to being enclosed and reserved for coppice.[52] What happened here can be paralleled in English iron-making districts such as the Weald or the Forest of Dean, which remain strikingly well-wooded to the present day. The danger to the Argyll woods returned when the charcoal iron industry finally gave up the struggle in the nineteenth century, and the oakwoods reverted to their modern overgrazed condition. But at least many of them are still there.

So it is the Irish case of desolated woods that most needs explaining. Some have suggested that there was less care for the woods, more an attitude of plunder, because Irish landownership seemed relatively insecure in the seventeenth century. Then there is the question of military advantage, well-expressed by the commentator in 1601 who described the thick Irish woods as 'a great hindrance to us and a help to the rebel, who can, with a few men kill many of ours . . . in the woods they may fall into an ambushcado'. What this writer proposed was not, actually, the extirpation of the woods, but their transformation into something like an English park, or perhaps coppice-with-standards, with a tree every twenty yards 'and the shrubs either stocked up at the first or continually cut up'.[53] But that there was a continuing military consideration is indicated, for example, in the 1660s when one of the partners at the Enniscorthy ironworks in County Wexford wrote:

> I verily believe that if there were not woodcutters, colliers etc. day and night in the woods, there would be meetings of Irish, and Tories would get together, the consequences whereof would peradventure be destroying of men, which is a greater evil than destroying woods.[54]

Surely, the major reason why the woods did not survive the charcoalers is a matter of opportunity cost. As the studies of Barnard have shown, iron manufacture in Ireland was of very marginal profitability. The ore had to

[52] J. M. Lindsay, 'Charcoal iron smelting and its fuel supply: the example of Lorn Furnace, Argyllshire, 1753–1876', *Journal of Historical Geography*, 1 (1975), pp. 283–98.
[53] Quoted by V. Hall, in Pilcher and Mac an tSaoir, *Woods, Trees*, p. 24.
[54] Quoted in T. C. Barnard, 'An Anglo-Irish enterprise: iron-making at Enniscorthy, Co. Wexford, 1657–92', in *Proceedings of the Royal Irish Academy*, 85, C (1985), p. 140.

be imported if iron of marketable quality was to be made. Wages of skilled workers were high because they too had to be tempted from England. On the other hand, the returns for tilled arable and good quality grazing land were respectable and offered a permanent means of keeping, on the estate, a rapidly growing population of people and a still more rapidly growing population of their animals. To fence a wood and lay it out in coppice rotation not only put up the cost of charcoal manufacture (at Enniscorthy already 25 per cent of the cost of making pig iron) but deprived the land of alternative use for as long as the ironworks persisted. So Irish landlords, even those like Sir William Petty in Kerry, who were alive to problems of timber supply in England, treated Irish ironworks as a way of making some money while clearing land for other purposes. Barnard finds him in 1672 considering 'how we may destroy the woods of Kiltankin and Kilcasheene to our best advantage, viz. how we may engage the iron works to take them off presently'.[55]

So it became characteristic of Irish ironworks that they perambulated from place to place, cutting down fresh woods as supplies of local fuel became exhausted. The woods did not re-grow because they were grubbed up or simply devoured by the multiplying peasant and his stock. By 1700, they were largely finished, and the Irish iron industry was largely finished with them. The reason why the iron industry just about hung on in Argyll, despite similar disadvantages of imported ore and high labour costs and competition from Lowland ironworks starting with Carron, is probably because there was not the same pressure for alternative land use.

This tells us, I think, that the history of land use has less to do with blind environmental wickedness than with the economic law of comparative advantage. Some, of course, would say that this was much the same thing, but at least it had its own rationality. The other lesson is that, if a society is sufficiently energy rich in most respects, it does not have to be too worried about all aspects of energy in order to become a highly successful economy. Scotland was certainly this before 1840, measured in growth of GNP per head; Ireland certainly also was successful, also before 1840, measured in the growth in the number of people supported (not a bad measurement if you consider contemporary priorities and avoid hindsight); Iceland equally certainly was entirely unsuccessful, unless you count mere survival a merit. I am not saying that energy wealth created success, but it was clearly impossible to do much without it.[56]

---

[55] T. C. Barnard, 'Sir William Petty as Kerry ironmaster', *Proceedings of the Royal Irish Academy*, 82, C (1982), p. 31.

[56] Richard Oram correctly points out that recent research gives a more nuanced and favourable view of early Icelandic fuel management than the picture given here. See his review in *Scottish Historical Review*, 90 (2011), pp. 172–3.

Fig. 8 Sir John Sinclair (1754–1835), Caithness landowner, improver, editor of the *Statistical Account of Scotland*, President of the first Board of Agriculture that organised the county *Reports on Agriculture* throughout Britain: 'the favourite object of my life, the collecting of useful information'. Portrait engraved by William Skelton, after Sir Thomas Lawrence, published 1790. © National Portrait Gallery, London.

CHAPTER 8

The Improvers and the Scottish Environment: Soils, Bogs and Woods*

By the first half of the eighteenth century, Scotland had been suffering from several thousand years of slow environmental degradation without any effective actions having been taken to reverse it, and from at least two centuries of declining and stagnant living standards, which flight and famine on a rare scale had failed to relieve. That did not make her exceptional among north-west European countries, except perhaps in the determination of so many of her inhabitants to seek their fortunes elsewhere by emigrating. In the course of the succeeding century, in the rural Lowlands, though perhaps not in the Highlands, environmental degradation had been counteracted, and the living standards of the population in the rural Lowlands, though evidently not in the Highlands, had improved. That fact too was not unique in north-west European countries, though Scotland and Denmark probably vied for having made most progress. The economic, social and political leaders of the countryside in this dramatic century were the Improvers, once much praised as enlightened patriots, now more subject to historical deflation. By the term Improver is to be understood those landowners of the period 1720–1820, assisted by their land stewards and large tenants, who, in their own districts, deliberately broke with the agricultural traditions of the past. One of the questions to be considered is whether their attitudes and actions contributed significantly to the turnaround. Did they address what the Danish historian Thorkild Kjærgaard describes for his country as an ecological crisis overcome by a green biotechnological revolution?[1] Or were they irrelevant?

My opening began with sweeping generalisations which demand some justification. A declining standard of living? Surely so, when most people's incomes consisted of food, and the most obvious change over the years

---

* This is a slightly amended version of a paper that first appeared in T. M. Devine and J. R. Young (eds), *Eighteenth Century Scotland: New Perspectives* (East Linton, 1999), pp. 210–24.
[1] T. Kjærgaard, *The Danish Revolution, 1500–1800: an Ecohistorical Interpretation* (Cambridge, 1994).

1500–1750 was from a country famous for eating animal produce to one famous for eating oatmeal – certainly plentiful and highly nutritious, yet Dr Johnson's snide observation about oats being horse-feed in England and human food in Scotland struck home.[2] The reason is clear enough: 500,000 people could live in Scotland enjoying an animal-based diet supplemented by meal, but 1,250,000 could not. When Rev. Dr Walker toured the Hebrides in the 1760s, he found the old-fashioned diet only on Jura, and partly on Rum, both with a high land-to-person ratio: his own calculations estimated the inhabitants of Jura as one per 246 acres ('a most melancholy Proportion'), of Rum one per 53 acres, against a Hebridean average of one per 30. On Jura, the people lived on milk, butter, cheese, fish, mutton, venison and very little vegetable food, and were 'in general a long lived people'. On Rum, he told of an old man who died aged 103 without having tasted bread until he was 50:

> This old man used frequently to remind the younger People of the simple and hardy Fare of former Times [i.e. animal food, fish and milk] ... and judged it unmanly in them to toil like Slaves with their Spades, for the Production of such an unnecessary Piece of Luxury [as bread].

The old man's perspective was entirely realistic, but Dr Walker was keener on high population than high standard of living.[3]

If you prefer statistics to this culinary fare, a comprehensive study has shown how, in Edinburgh, real wages, measured against oatmeal prices, fell by 25 per cent for masons and 17 per cent for labourers between the mid-sixteenth century and the mid seventeenth, and did not obviously improve there or anywhere else in Scotland again for another hundred years or more; this was in marked contrast to the situation after 1650 in England.[4] As for flight and famine on a rare scale failing to relieve the situation, estimates for emigration suggest that (taking together the flight to Ulster, Poland and the mercenary armies) in the first half of the seventeenth century a fifth of the marriageable males may have left the country, and that overall for the whole century the emigration rate was probably the highest in Europe.[5]

[2] A. J. S. Gibson and T. C. Smout, 'Scottish food and Scottish history, 1500–1800', in R. A. Houston and I. Whyte (eds), *Scottish Society, 1500–1800* (Cambridge, 1989), pp. 59–84.

[3] M. M. McKay (ed.), *The Rev. Dr. John Walker's Report on the Hebrides of 1764 and 1771* (Edinburgh, 1980), pp. 115, 196.

[4] A. J. S. Gibson and T. C. Smout, *Prices, Food and Wages in Scotland, 1550–1780* (Cambridge, 1995), chs 8 and 9.

[5] T. C. Smout, N. C. Landsman and T. M. Devine, 'Scottish emigration in the seventeenth and eighteenth centuries', in N. Canny (ed.), *Europeans on the Move: Studies in European Migration, 1500–1800* (Oxford, 1994), pp. 76–90.

Such data as we have on famine suggests that in the 1690s the population of Scotland might have fallen by 13 per cent, though not only by mortality.[6] Neither set of events was followed by a rise in real wages or a clear improvement in the standard of living, suggesting that the marginal productivity of labour remained close to zero. The only time that wage rates showed a jump and English standards seemed within reach was, ephemerally, in the Cromwellian 1650s, at the end of the disruption, bloodletting and plague of civil war; but those good days for survivors ebbed within a decade.[7] Revisionist historians who seek to put the brightening days of Scottish economic history decisively back to the seventeenth century need to come up with a better supported case than isolated examples of good practice and wealth generated by a handful of merchants, usually abroad.

How about the alarming generalisation about those thousands of years of environmental degradation accompanied by inaction? This hinges on the long-term availability of plant nutrients, in the face of change induced by a combination of natural and anthropogenic forces. When the glaciers of the last Ice Age disappeared 10,000 years ago, they ground the rocks into a highly nutritious mineral mix, and in due course the face of most of Scotland became clothed with forest. For farmers, the crucial element is nitrogen in a form available to plants, a chemical element particularly liable to diffusion into the atmosphere, but which forests are good at capturing. The main source of fertility available to Neolithic, Bronze-Age and Iron-Age farmers was the high quality nitrogenous humus on cleared forest land. Most of the woods on cultivable land had probably been cleared at least once before the Roman invasion (remember they can regrow, on a slash-and-burn regime common in parts of France or Finland into the eighteenth or even early nineteenth centuries). Although the extent of woodland cover ebbed and flowed subsequently, in response to population fluctuation and climatic change, the trend was always downwards. Archaeologists will probably never completely satisfy themselves about the relative contributions of human activity and climatic change in the downfall of the primal woods and the increase of bogs, but from earliest times they must have been mutually reinforcing. The Scottish Lowlands at least were finally, largely, though never totally, denuded of wood by the end of the thirteenth century, a decline in woodland cover from possibly a prehistoric maximum of at least 50 per cent to, at most, 10 per cent. The situation in the Highlands probably left more woodland cover than the average, but that on

---

[6] R. E. Tyson, 'Contrasting régimes: population growth in Ireland and Scotland during the eighteenth century', in S. H. Connolly, R. A. Houston and R. J. Morris (eds), *Conflict, Identity and Economic Development: Ireland and Scotland, 1600–1939* (Preston, 1995), p. 67.

[7] Gibson and Smout, *Prices*, pp. 275–6.

difficult soils. One would have expected woodland recovery in the century and a half following the demographic catastrophe of the Black Death and associated events, but it is hard to find, possibly because increased cold and wet discouraged it, possibly because extensive grazing did the same, probably because of a fell combination of the two. These were the centuries of maximum complaint about wolves – fewer people, lots of sheep, cattle and horses, a lupine paradise.[8]

The gradual and fluctuating trend of woodland decline over a period of 5,000 years does not detract from its serious character or its fatally debilitating nature. The removal of forest cover starts a chain reaction. More water is released into streams and rivers, because the ability of the trees to return it to the atmosphere through their leaves, or to delay its course through the soil by their roots, disappears. From this follows erosion, more rapid leaching of mineral nutrients left by the grinding glaciers, more rapid dissipation of nitrogen, acidification of the soil, formation of podsols and other alterations of soil structure associated with water logging, floods, bog formation, and so forth. Such deterioration, especially rising water levels, is likely to kill other existing woods, and to press farmers on to partially elevated lands, the slopes of the valleys, where inputs of nitrogen and other nutrients through manure are likely to wash out even more rapidly. As the Englishman Thomas Morer put it in 1689:

> They have many fine vallies . . . almost useless, on the account of frequent bogs and waters in such places . . . [yet] 'tis almost incredible how much of the mountains they plough, where the declensions, I had almost said the precipices, are such, that to our thinking, it puts 'em to a greater difficulty and charge to carry on their work, than they need be at in draining the vallies.[9]

Morer's solution showed little understanding of the problems of drainage in Scotland, but there are many independent witnesses to his picture of fertile ground under water and marginal land in cultivation.

Environmental change may be almost imperceptible from one generation to another, but it is change nevertheless. Enhanced rainfall, from whatever cause, steadily lowers the pH value of the soil as it washes out the

---

[8] For Scottish woodland history, see, for the prehistoric, R. Tipping, 'The form and fate of Scotland's woodlands', *Proceedings of the Society of Antiquaries of Scotland*, 124 (1994), pp. 1–54; for the Middle Ages, there is an indication in J. M. Gilbert, *Hunting and Hunting Reserves in Medieval Scotland* (Edinburgh, 1979). The figures for woodland cover are my own estimates.

[9] P. H. Brown, *Early Travellers in Scotland* (Edinburgh, 1891), pp. 266–7.

natural lime; if the pH level falls from 7 to 4, this inhibits the growth of nearly all agricultural plants, and, even more seriously perhaps, stops the growth of wild clover and other leguminous plants with nitrogen-fixing powers. Once the forest humus is gone (and its initial nitrogen returns to the atmosphere relatively quickly), a problem arises. The other main sources for nitrogen before the agrarian changes of the eighteenth century were manures, mainly dung, locally seaweed or turf, and the atmosphere itself. Calculations concerning eighteenth-century Denmark suggest that farmers could contribute to cultivated ground about 30 kilogrammes of nitrogen per hectare per year from the middens, equivalent to only 10–15 per cent of the nitrogen supplement given on Danish farms in the 1980s, and that on uncultivated grazing ground about 20 kilogrammes fell from the atmosphere or was produced by wild plants. This was the basis of contemporary low yields.[10]

But – here is the gist of the matter – nitrogen levels were not only low, but by the logic of rainfall's eternal leaching of minerals, yields were prone imperceptibly to fall forever unless an increasing input of manure could somehow be provided by ever-increasing human effort in the manufacturing of soils. From the thirteenth century onwards, the construction of deepened soils spread throughout north-west Europe, the so-called plaggen soils of the Netherlands, Germany and Ireland, which have also been identified in Scotland. They were made by transporting volumes of turf from the hill to the inbye land, impoverishing the hill and enriching the inbye, an enormously labour-intensive task which, after 800 years, provided a topsoil of eighty centimetres depth at Papa Stour in Shetland. Such labour constructed phosphorous-rich soils, but the nitrogen bonus from them was short lived and depended on constant laborious replenishment from new turf.[11]

The great cornucopia of nature, in fact, had a slow leak. It is no wonder that the archaeologists at Skara Brae, excavating a Neolithic farming and hunter-gathering community at the very start of this process, marvel at the plentiful resources and apparently easy life enjoyed by prehistoric man, compared to his early modern counterpart.

Seventeenth-century Scots (like seventeenth-century Danes) had no response to this conundrum. Little that was done by way of agrarian initiative then addressed these problems: consolidation of holdings and longer leases probably made the mobilisation of labour easier, to hold the problem at bay for a moment, but did not address the root cause of the matter. The

[10] Kjærgaard, *Danish Revolution*, p. 22.
[11] D. A. Davidson and I. A. Simpson, 'Soils and landscape history: case studies from the Northern Isles of Scotland', in S. Foster and T. C. Smout (eds), *The History of Soils and Field Systems* (Aberdeen, 1994), pp. 66–74.

slow and partial introduction of liming, for example in the Lothians, Fife and Stirlingshire, was more relevant, as where it was practised it visibly increased yields by changing the pH level. Perhaps the only widely popular and effective Scottish response was to go to Ulster. The coincidence is striking of the prosperous century for the farmers there compared to the dismal years just over the Irish Sea, especially in the 1690s. It cannot conceivably be due to any difference in climate. Was it because Ulster, when invaded, was among the most forested parts of Ireland, and the settlers found a nitrogenous humus-bonus just as the English did in North America? And was the subsequent migration of the Scots Irish, in the decades hinging on the 1720s to Pennsylvania and the Carolinas, related to the search for another humus bonus as the Ulster one became exhausted? It merits investigation.

So at last we come to the Improvers. Did they have remedies for the ecological degradation we have described? The question does not demand that they understood the scientific reasons behind the problems; plainly, in a pre-scientific age they could not. Nor does it demand that their remedies, if they had any, should be immediately adopted by the entire farming community. Nor does it demand that they should make a profit from their ideas, or even avoid bankruptcy in their efforts to carry them out. But it does demand that they had, from empirical observation, some correct notion that their ideas would transform the productive capacity of the land, and it also demands that, ultimately, someone, not necessarily themselves, should carry them forward. Let us examine these questions by looking firstly at soils, secondly at bogs, and thirdly at woods.

The Improvers' attitudes to soils should have concentrated upon two things: legumes and lime. Red clover was the best of the legumes, a family of plants that almost alone could call down nitrogen from the air and fix it in the soil. Peas, beans, sainfoin and wild white clover belonged to this family, and could do the same as red clover to a more limited extent; of these, peas and beans had been widely cultivated in the Lowlands since the Middle Ages. A range of historians, but especially C. P. H. Chorley, have, however, singled out the introduction of clover as overwhelmingly the most important cause of the increase in agricultural productivity between 1750 and 1850, by which time Peruvian guano and later Chilean nitrates were coming to the aid of the nitrogen deficit.[12] In the Scottish context, enthusiasm for clover is usually disguised under a general enthusiasm for sown

---

[12] C. P. H. Chorley, 'The agricultural revolution in Northern Europe, 1750–1880: nitrogen, legumes and crop productivity', *Economic History Review*, 34 (1981), pp. 71–93. See also Kjærgaard, *Danish Revolution*, pp. 84–7.

grasses and rotations that alternated 'green' crops with 'white' – i.e. legumes with corn. Characteristically, in East Lothian, this entailed a six-course rotation of wheat, peas and beans mixed, barley, sown grass, oats and fallow. By sown grasses was generally understood clover and rye grass. The key figure here was the sixth Earl of Haddington, whose *Short Treatise on Forest Trees, Aquatics, Evergreens, Fences and Grass Seeds* (Edinburgh, 1735), concludes with a discussion on the relative 'utility of sowing bread clover, small white clover, mixtures of clover and rye grass, or rye grass by itself, sanfoin and lucerne'.[13] There is little doubt of the importance attached to sown grasses, or the speed with which clover spread in the decades before 1790. In East Lothian, the minister of Whittingehame in the *Statistical Account* said 'a trifling quantity of clover and rye grass seeds might be sown, upwards of 50 years ago; but it will not exceed 37 years, since it became the universal practice among farmers'. He added that 'turnips and grass crops are certainly among the greatest and most valuable improvements, which have been made in agriculture', a comment echoed by his colleague at Athelstaneford, who listed the Earl of Haddington's introduction of clover among 'the first, and will probably prove to be the most important and permanent, of all improvements in modern husbandry'.[14] The purist will correctly want to object that Haddington may not deserve his reputation for being the first to introduce clover, and to point out that it was sometimes grown earlier on other estates. But the prestige and success of his example undoubtedly helped to push it on its way. By the 1790s, 46 per cent of parishes in Lanarkshire, 62 per cent in Fife and 71 per cent in Angus were using sown grass rotations.[15]

But if the Improvers were keen on clover, they were no less keen on liming, either through the application of marl dug from calcareous deposits, or through the distribution of lime made from limestone in kilns throughout the country. Here one could certainly not award eighteenth-century Improvers the priority of discovery or introduction: lime was well known in the seventeenth century as the usual way of improving land in the east Central Belt, and had been used on grass in Ayrshire since at least the 1590s.[16] However, the area using lime before 1700 appears to have been comparatively restricted, and the Improvers universalised its use across

---

[13] J. E. Handley, *Scottish Farming in the Eighteenth Century* (London, 1953), pp. 145–6.

[14] *The Statistical Account of Scotland* (henceforth, *OSA*) (Wakefield, 1972–83), vol. 2 (*Lothians*), pp. 649, 452.

[15] T. M. Devine, *The Transformation of Rural Scotland: Social Change and the Agrarian Economy, 1660–1814* (Edinburgh, 1994), p. 54.

[16] Ian Whyte, *Agriculture and Society in Seventeenth-Century Scotland* (Edinburgh, 1979), pp. 198–208.

Lowland Scotland. The most remarkable example of this was the Earl of Elgin's £14,000 investment at Charlestown in Fife in 1777 and 1778: he built nine large draw-kilns, a harbour, waggon-ways and a village for the 200 employees who quarried the 80–90,000 tons of limestone and loaded the slaked lime onto 1,300 separate cargoes a year, with a total annual value of over £10,000. The concern burnt some 12,000 tons of coal a year, and the lime was distributed throughout the Forth and the Tay, and as far as the north of Scotland, with between thirty and fifty ships queuing in summer at Charlestown to load their cargoes.[17] In other areas – for example in Dumfriesshire, and near Roscobie Loch in Angus – the digging of shell marl provided similar benefits. These were instantly recognised, though not understood, when they changed the pH of acid ground. Thus Robertson's account of south Perthshire agriculture in 1794:

> There is nothing more common, and perhaps few things more diffi-cult to be accounted for, than, when lime is spread on short heath, or other barren ground, which has a dry bottom, to see white clover, and daisies, rising spontaneously and plentifully, the second or third spring thereafer, where not a vestige of either, nor even a blade of grass, was to be seen before.[18]

There was a third novelty of improving agriculture, which we do not propose discussing, as its main impact came in the nineteenth century, beyond our immediate purview: this is the introduction of turnip hus-bandry. Before 1800 it had made relatively minor progress beyond Fife, Angus and East Lothian, though where it had been introduced it also greatly helped the nitrogen crisis by allowing more of those dung-producing machines, cattle and sheep, to be kept through the winter. The fourth novelty of great importance was the potato, so immediately accepted that the Improvers are usually overlooked as those who urged its adoption in the first place. We shall not deal with that here, as it was more of a crop novelty than something that had an impact on soil as such. But, of course, in a Boserupian way, it provided an effective response to the conundrum of rising population in a still deteriorating Highland environment.

In other ways the Improvers' attitude to soil was much more conserva-tive; the emphasis on legumes and lime was genuinely innovative, but much of the Improvers' effort went into the old, uphill task of manufacturing soil

---

[17] *OSA*, vol. 10 (*Fife*), pp. 308–9.
[18] J. Robertson, *General View of the Agriculture in the Southern Districts of the County of Perth* (Edinburgh, 1794), p. 26.

with muck and turf. It was not entirely futile, though some of the effort was questionable. The use of human dung on the fields within about five miles of Edinburgh enabled large increases in production of grain, especially of wheat, and similar benefits resulted near other towns. Price movements indicated the growing appreciation of town ordure: according to the minister of Duddingston, the price of a cart-load of the offals of the streets of Edinburgh had risen from 2d. to as high as 1s. 6d. between 1730 and the 1790s. It normally went free on the toll road, though an attempt by the turnpike operators round Edinburgh to charge for it led in 1787 to a boycott – 'no farmer would purchase it, so that the good town had to collect the fulzie itself for several weeks, till the husbandmen came into better temper'.[19]

That, however, was quite local, and there were many other local solutions to the nitrogen problem, mostly either opportunistic or involving large labour inputs to provide short-term benefits, in the same way as the age-old manufacture of plaggan soils. It did not need Improvers to teach the use of kelp-spreading on the field, within reach of the coast; in some places mussels were similarly used as manure by the local people. More illustrative of the restless innovation and opportunism of the Improvers was the activity of John Auldjo, a merchant of Aberdeen, who manured his ground with a mixture of horse dung, spent tanbark and composted dogfish bought in quantity from local merchants after the fish livers had been extracted for the oil; in Ayrshire, farmers used soap waste, in Clydesdale they mixed midden refuse with horn shavings, soot, woollen rags, parings of leather, peat and coal, and in Aberdeenshire in 1811 they used whale blubber from the Greenland fleet, which raised 'a very rich crop for the first year, and a tolerable one for the second'.[20]

If it is difficult to imagine a sustainable improvement in agricultural productivity based on such hand-to-mouth resources as dogfish waste and whale blubber, an even bigger question mark hung over the publicised techniques of paring and burning turf. In a way, this was plaggan extended. The aforementioned Auldjo of Aberdeen, for example, had his own recipe: he took a large quantity of turf from the moss, set it alight, added a layer of clay, another of turf, another of clay until the pile was as high as a man could reach, getting 1,000 cart-loads at a single burning. According to Alexander

---

[19] OSA, vol. 2 (Lothians); G. Robertson, General View of the Agriculture of the County of Midlothian (Edinburgh, 1793), pp. 48–9.

[20] A. Wight, Present State of Husbandry in Scotland (Edinburgh, 1778–84), vol. 3, pp. 599–600; W. Fullarton, General View of the Agriculture of the County of Ayr (Edinburgh, 1794), p. 52; G. S. Keith, General View of the Agriculture of Aberdeenshire (Edinburgh, 1811), p. 432.

Wight, sixty loads on an acre of oats or peas did wonders, and thirty to forty loads as a top dressing for grass 'never fail to mend the crop'.[21]

Most commentators in the period of the *General Views of Agriculture,* 1794–1814, were against the practice of burning, however: as was said of Ayrshire, 'by burning moss you may raise one great and one moderate crop, but the moss will never produce another of much value', and in Roxburghshire 'it reduces to ashes some of the best soil, and has hurt ten acres for every one it has benefited'. Others were not so negative, recommending paring and burning to bring into cultivation the wide wastes of Inverness-shire, or as the best method of improving deep, peaty land in Perthshire. Yet advocacy of this technique scarcely shows the Improvers in a very progressive light. Paring and burning was, firstly, very labour intensive; secondly, provided ephemeral benefits in respect of the nitrogen deficit; thirdly, actually impoverished the moor and accentuated the difference between inbye and outbye.[22] In popularising clover and greatly extending liming, the Improvers provided things qualitatively new and critical; in other directions they were often drawing experimental blanks or trying to breathe new life into ancient desperate remedies.

The question of bogs illustrates something rather different: the ability of the Improvers to take a completely new look at a natural resource. As Tom Devine puts it, 'The rationalism of the Enlightenment helped to change man's relationship to his environment. No longer was nature accepted as given and pre-ordained; instead it could be altered for the better by rational, systematic and planned intervention.'[23] A reading of Macfarlane's *Geographical Collections,* mainly composed of observations between about 1600 and 1730, gives an overwhelming impression of nature as given and preordained, and often appreciatively described. Bogs were seen as a useful resource and a blessing to the community, a place to dig fuel, perhaps also to trade in it to neighbouring downs, and a source of 'bog fir' for building. Conversely, a parish without such resources was cursed with a singular misfortune of want of peat.[24]

By the age of the Improvers, however, bogs came to be viewed as scandalous wastes and opportunities for a human response to the challenge. Thus

---

[21] Wight, *Present State*, p. 600.
[22] W. Aiton, *General View of the Agriculture of the County of Ayr* (Edinburgh, 1811), p. 374; R. Douglas, *General View of the Agriculture of the Counties of Roxburgh and Selkirk* (Edinburgh, 1798), p. 132; J. Robertson, *General View of the Agriculture of the County of Inverness* (Edinburgh, 1808), p. 231; J. Robertson, *General View of the Agriculture of the County of Perth* (Edinburgh, 1799), p. 274.
[23] Devine, *Transformation*, p. 65.
[24] A. Mitchell (ed.), *Geographical Collections Relating to Scotland, Made by Walter Macfarlane* (Scottish History Society, Edinburgh, 1906). See also Chapter 6.

Andrew Steele, in the opening to a treatise on peat moss in 1826, spoke of the bogs as 'immense deserts', and Robert Rennie in 1807 of 'innumerable millions of acres [that] lie as a useless waste'.[25] Those who had got to grips with the bogs were heroes and patriots, like Lord Kames and particularly his son-in-law, whose conquest of Blairdrummond Moss so fired the contemporary imagination. William Aiton, in 1811, compared Lord Kames with David Dale, the cotton manufacturer, much to the advantage of the former. Both had provided a public benefit by employing displaced Highlanders, but whereas at Blairdrummond 'the moss colony remain healthy and happy, delighting in their situations, warmly attached to their patron, and to the Government', the Highlanders at New Lanark 'became discontented' and left.[26] Soon most of Kames' neighbours were following the family example: within a century from 1766 an area of sixty square kilometres of the Stirlingshire carse had been cleared of peat moss and heather to a depth of up to four metres. This phenomenal quantity of waste was thrown into the Forth and floated out to sea, where it damaged fisheries and oyster beds some fifty kilometres east of the source of the pollution.[27]

There is no doubt that reclamation of bogs, large and small, ultimately provided many hundreds of thousands of acres for agriculture, and in one sense rolled back the consequences of millennia of environmental impoverishment of a natural or anthropogenic character – until late in the present century people began to worry if any of the characteristic biodiversity of the peatlands would remain and consequently put the brakes on the process. Probably even more important, however, was the general boost that the reclamation campaign gave to field and land drainage. This culminated in the 1830s in the enthusiasm for subsoil ploughing and tile drainage, above all associated with Smith of Deanston, innovations which were able to 'turn sour, waterlogged soil into high quality, productive land . . . [so that] the change must have seemed nearly miraculous'.[28] Whereas in the past, in the words of the minister of the Aberdeenshire parish of Tarves, 'the stagnation of water on the low ground utterly precluded tillage, while the arable lands were . . . chilled from November to May by innumerable land springs',[29] now high water levels that had persisted, perhaps, for millennia since

[25] Steele, *The Natural and Agricultural History of Peat-moss or Turf-bog* (Edinburgh, 1826), p. 1; R. Rennie, *Essays on the Natural History and Origin of Peat Moss* (Edinburgh, 1807).

[26] W. Aiton, *A Treatise on the Origin, Qualities and Cultivation of Moss-Earth, with Directions for Converting it into Manure* (Ayr, 1811).

[27] D. S. McLusky, 'Ecology of the Forth estuary', *Forth Naturalist and Historian*, 3 (1978), pp. 10–23.

[28] A. Fenton, *Scottish Country Life* (Edinburgh, 1976), pp. 21–2.

[29] Francis Knox, writing in the *New Statistical Account* (Edinburgh, 1842) and cited by W. A. Porter, *Tarves Lang Syne, the Story of a Scottish Parish* (York, 1996), p. 36.

deforestation, dropped. Consequently, minerals in soils previously unavailable to crops on account of the 'slow biological turnover' became part of the farmers' resource.[30]

Drainage schemes were all expensive, and early pioneers often over-reached themselves. However, when Thomas Hope of Rankeillor, president of 'The Honourable Society of Improvers' from 1723, drained Straiton's Loch on what became the Meadows of the south side of Edinburgh, to demonstrate to his fellow landowners what could be done with boggy land, he was also demonstrating the mind-set of the Enlightenment in respect to the possibility of altering the inherited environment.[31] There is a tower of effort, stone patiently added upon stone by trial and error, between the tentative experimental optimism of Hope of Rankeillor and the well-grounded opinions of Smith of Deanston a century later. That so many of the early Improvers, like Hope, lost money, no more shows the futility of their efforts than the financial problems of Sir Clive Sinclair shows that he should not have been inventing calculators or electric cars.

Our third examination is into Improvers and woods. Probably because of their unusual combination of beauty and utility, tree-planting programmes around the policies of landowners were already well established in the seventeenth century, but it took the Improvers to begin planting on a large scale and away from the immediate surroundings of the great house. Some of the great eighteenth-century names – Grant of Monymusk, the Earl of Haddington, the Duke of Atholl – planted millions of trees in their lifetimes, some on new sites, like Haddington's Binning Wood on a sandy waste near Tyninghame, some on old sites, like the Earl of Moray's replanting with beech and other trees the site of the ancient Forest of Darnaway.[32] Frequently trees were planted as windbreaks on the field margins to create a more favourable micro-climate: the upland edges of Lothian and the Borders are full of lines and spinnies of beech and sycamore planted to improve the chances of crops and sown grasses higher up the hill. For the first time, deliberate anthropogenic reforestation was replacing deforestation, not because Improvers had a detailed knowledge of ecological science but because, firstly, wood was the source of several essential raw materials in diminishing supply in an expanding economy; secondly, trees were ornamental and prestigious (a well-planted estate was considered well-cared-for) and thirdly, it was perceived by empirical observation that they protected and 'warmed' the land. The Fourth Duke of Atholl, known to his

[30] Kjærgaard, *Danish Revolution*, p. 57.
[31] J. A. Symon, *Scottish Farming Past and Present* (Edinburgh, 1959), p. 302.
[32] See M. L. Anderson, *A History of Scottish Forestry* (Edinburgh, 1967), vol. 1, ch. 10.

contemporaries as 'Planter John', wrote of his motives in his private forestry diary: 'In my opinion, Planting ought to be carried on for Beauty, Effect and Profit', putting the financial motive third. Beauty and effect equated to reputation and power, and though planting had an aspect of conspicuous consumption, it was not, in any nobleman's mind, to be equated with waste. A duke or an earl needed to be seen to embellish his estates. In the case of the dukes of Atholl, who, from the early eighteenth century to 1830, planted over 21 million trees on 15,000 acres of ground, there was a considerable degree of public benefit. They were the first to introduce the larch as a commercial crop excellent for ship construction, house-building and fencing, and the Fourth Duke estimated that he could supply the Royal Navy with most of their needs – though this came to nothing.[33] In absolute terms, the Improvers cannot have added much to the tree cover of Scotland – perhaps 2 per cent of the land surface was re-afforested by them in the eighteenth century. But they made a start.

The Improvers, on balance, also introduced very much better standards of care for the semi-natural woods, which were the precious descendants of the original forest cover.[34] Certainly the situation varied from place to place and from time to time; few lairds who found themselves suddenly assailed by financial hardships could resist making depredations on their woods as a way of trying to meet them – this was true of families as various as the Dukes of Queensberry and the Grants of Rothiemurchus. On the other hand the damage done by such forays could be restored – as indeed at both Drumlanrig and at Rothiemurchus – by good stewardship plus natural regeneration in the next generation of the family, so that clearfelling by no means necessarily meant the demise of a wood.

To some extent, what happened to the pinewoods may have depended on which side of Scotland they stood. Quite severe felling episodes in the Spey woods (Abernethy and Rothiemurchus), or those along the Dee (Glentanar) or Carron (Amat) in eastern Scotland were generally followed by regeneration, but in the west, felling at Glen Orchy, Glencoe and Loch Etive led largely to extinction. There is possibly a climatic element at work – in the west, heavy moss cover may have made regeneration impossible after felling in the eighteenth century, due to climatic alteration over the period since the standing generation of trees were seedlings, up to 300 years earlier. Improving noblemen like the Earls of Breadalbane and the Dukes of Argyle presided over this destruction, which Breadalbane at least noticed

---

[33] S. House and C. Dingwall, '"A Nation of Planters": introducing the new trees, 1650–1900', in T. C. Smout (ed.), *People and Woods in Scotland* (Edinburgh, 2003), pp. 135–42.

[34] T. C. Smout, A. R. MacDonald and F. Watson, *A History of the Native Woodlands of Scotland, 1500–1920* (Edinburgh, 2005). See also Chapter 3 above.

in retrospect with dismay, but they had no remedies. Enclosure of the pine-woods to keep out peasant stock was apparently introduced as a novel tool in pinewood management by the Committee for Annexed Estates; it may have had some effect at the Black Wood of Rannoch, but was not always contin-ued for long enough to make a difference.[35] Other improving landowners, especially the Earls of Fife in the Forest of Mar, effectively prevented regen-eration in the woods (despite their commercial value) by encouraging deer for sporting purposes at the end of the eighteenth century, enclosing them *inside* the forests. The parlous state of the upper Deeside Caledonian pine at present is the result of 200 years or more of systematic misuse, inaugurated mainly by the Fife family.[36]

Nevertheless, the Improvers must be credited with having done a great deal to halt the decline of broad-leafed woodland, especially the natural oaks of the west coast and Perthshire, when the market for charcoal and tanbark dramatically improved from the middle of the eighteenth century. It is true that some of the earliest ironmasters may have inflicted damage where landowners failed to check them and when grazing of the regen-eration was allowed to follow cutting – this is suspected, for example, around Loch Maree and in Glen Kinglass. But from the 1750s, when the Lorn Furnace Company arrived in Argyll, ironmaster and landowner alike co-operated to improve the care of the woods, led by the Duke of Argyle. They did this by fencing, systematic use of coppice, and exclusion of peasant stock, especially by banning the keeping of goats.[37] Tanbark by 1800 was providing a better reason for good woodmanship than charcoal manufacture, even in the west; in Perthshire and Menteith it had always been tanbark that was the most important woodland product. The owners of tanbark woods, like the Dukes of Montrose along Loch Lomondside and in the Trossachs, and the Dukes of Atholl at Dunkeld and elsewhere, and many others, similarly enclosed, coppiced, and excluded the peasants' animals. They also sowed a great many acorns, often on previously arable land, and weeded out less-valuable trees like birch, alder and willow to try to create a virtual monoculture of oak.[38]

All this was good practice unless you happened to be a tenant of one of these Improvers and relied on the woods to shelter your animals in winter.

[35] J. M. Lindsay, 'The use of woodland in Argyleshire and Perthshire between 1650 and 1850', unpublished University of Edinburgh Ph.D. thesis, p. 291.

[36] A. Watson, 'Eighteenth century deer numbers and pine regeneration near Braemar, Scotland', *Biological Conservation*, 25 (1983), pp. 289–305.

[37] J. M. Lindsay, 'Charcoal iron smelting and its fuel supply: the example of Lorn furnace, Argyllshire, 1753–1876', *Journal of Historical Geography*, 1 (1975), pp. 283–98.

[38] J. M. Lindsay, *thesis, passim*; R. M. Tittensor, 'History of the Loch Lomond oakwoods', *Scottish Forestry*, 24 (1970), pp. 100–18.

Their exclusion might occasion great hardship, and further trouble was occasioned by shortage of accessible wood for peasant use in the midst of natural plenty. In southern Argyll, for example, where farm houses had been bigger than average because of the abundance of oak, tenants were by 1798 having to import Norwegian softwood for their homes, as the woods were closed to them by the landowners.[39] Sometimes enclosure led to temporary rent reductions. More often, the lack of any consideration that tenants might need to be compensated for losses of this kind appears as character-istically high-handed behaviour by the Improvers. This compares badly with the situation in Denmark where a statute of 1805 established the right and duty of landowners to enclose the woods, but awarded tenants an equivalent area of grazing for the value of the rights they had lost by enclosure.[40] The penalty was seen later in the nineteenth century, when the value of tanbark and charcoal collapsed, lasting equivalent uses for oak were not found, the enclosures tumbled down and the woods reverted to grazing. As far as the farmers were concerned, that was all they were good for. Any lingering tra-dition of peasant care for the wood had evaporated, especially as there was less need for native wood as a raw material with the coming of cheap iron and steel for tools and the availability of foreign softwood for construction, thanks to free trade, steam ships and the railways.

To be fair to the landowners, it is nevertheless rather questionable how deep a tradition of peasant care for the woods there had ever been in Scotland. Perhaps this was because, however freely the tenants had helped themselves to supplies of bark and of building and tool timber in the past, it had always been clear in Scots law (in contrast, say, to Norwegian law) that the woods and the trees in them belonged to the landlords, not to the tenants, so they had no special incentive to treat them well.

Stories of the decay of the native woods, and serious economic and ecological consequences following, come in the late eighteenth and early nineteenth centuries from Sutherland and Wester Ross, where much of the wood was birch of little commercial worth, and where the reach of the market to enhance values for such oak as there was, was less significant than further south. Here one hears of straths becoming deforested for causes that baffled contemporaries, but which are less puzzling when one learns that local farmers kept up to several score of goats apiece, as well as cows and horses. One consequence of the decay of the woods was the need to keep cattle indoors for three or four extra months in the year, with greater mortality among the animals. Also, bracken and heather took over

---

[39] J. Smith, *General View of the Agriculture of the County of Argyle* (Edinburgh, 1798), p. 132.
[40] Kjærgaard, *Danish Revolution*, pp. 111–12.

the good grass sward that had grown beneath the trees, so the land was less able to support stock, though it was alleged that sheep could thrive on such degraded ground better than cattle. The reporter used this evidence of ecological decay (which was independently corroborated by Rev. William Sage, a man by no means sympathetic to the policy of clearance) as a justification for replacing cattle by sheep in the straths.[41]

Attention has been drawn before to ecological degradation thought to have been brought about by the impact of these post-clearance sheep on the uplands of the far north by about 1870; it is interesting to find that such degradation was apparently beginning to be noticeable well before the clearances.[42] The Sutherland estate, and in particular the arch-clearer Patrick Sellar, actually went to some trouble to preserve the surviving woods. Sheep farmers were instructed on pain of heavy fines not to allow their animals to 'injure or damage any tree or trees, or any of the natural woods', not to cut wood for their own purposes, and to use 'best endeavour, to protect, and increase the growth of the said natural woods'. Attempts at enclosure were made, then given up again as the cost was too high in wintering ground for the sheep, but careful management and control of the flocks grazing in the wood was insisted upon. Apparently, it worked within a generation. In 1830, the birches were said to be 'rising rapidly all over Assynt' and both there and in Strathnavar and Kildonan they needed thinning to bring the regeneration under control. In 1842, Patrick Sellar commented on the birch woods of Strathnavar as 'quite Choked with Willow and hazel', but they could not possibly be so described now.[43] The decline of sheep-farm profitability, followed by the decline of shepherding, are primarily to blame for the present parlous state of many of the Sutherland woods, not the introduction of sheep-farming itself.

So where does all this leave us? The Improvers were not people with particularly original minds, and all too often not people of a practical bent, as their frequent bankruptcy shows. They could never have improved Scotland unless price movements had rewarded their efforts after 1760, and particularly after 1790, so the demand side was critical. All they did was to help force the pace. Growing clover was already common in parts of England and the Netherlands, reclaiming land especially common in the Netherlands,

[41] J. Henderson, *General View of the Agriculture of the County of Sutherland* (London, 1812), pp. 83, 86, 106, 176–7.

[42] J. Hunter, 'Sheep and deer: Highland sheep farming 1850–1900', *Northern Scotland*, 1 (1973), pp. 199–222.

[43] R. J. Adam (ed.), *Papers on Sutherland Estate Management, 1802–1816* (Scottish History Society, Edinburgh, 1972), vol. 1, pp. 187–8, 197; vol. 2, pp. 149–50; M. Bangor-Jones, 'Native woodland management in Sutherland: the documentary evidence', *Scottish Woodland History Discussion Group Notes*, 7 (2002), pp. 1–4.

good forest practice common in England, Germany and France and perhaps also in parts of Lowland Scotland. But the cumulative effect of their stimulus, perhaps especially in soil treatment, was, indeed, improvement of an environment heavily stressed by long misuse. It would be nice if we could be sure that we were doing as well today as they were doing at the close of the eighteenth century.

Fig. 9 'St Mary's College and Queen Mary's Tree, St Andrews University' – a Valentine postcard of 1888. This seems to be the first reference to the old thorn in the quadrangle as having been planted by Mary Queen of Scots. It appears, like many similar, to be an imaginary association. The tree is probably no more than 300 years old. Photograph: Valentine Collection. Courtesy of St Andrews University Library: JV-9651.

CHAPTER 9

Trees as Historic Landscapes: from Wallace's Oak to Reforesting Scotland*

The main purpose of this paper is to sketch, in a Scottish context, the relationship between trees and our sense of landscape, especially of historic or cultural landscape. Trees have always been many things to many people, and it is as well to start by reminding ourselves that mostly they have been, for those who planted or managed them, timber: utilitarian, not ornamental.

Before the present century, Scottish wood was widely used when there were no cost-effective substitutes in the form of alternative materials or cheap imports. There was no subsidy or protection except that afforded by primitive technologies and underdeveloped markets, so when something better or cheaper was available, the woods were abandoned. Take its use as fuel. Where there was abundant coal or peat for domestic heating and industrial uses like salt panning, the woods were readily and anciently converted into arable or pasture land because there was no incentive to maintain them as a source of energy. The wide and desolate spaces of Buchan, or the empty fertile plains of Lothian, were described by seventeenth-century commentators as having an abundance of fuel that greatly added to the comfort of the inhabitants, but it was not wood. One of the reasons why Scotland has so little natural wood remaining compared to most European countries is just because it has so much fossil or semi-fossil fuel. Fuel-providing woods survived only where they had a specialist use: birch wood was unbeatable for smoking fish, and coppiced for that purpose in the hinterlands of fisher towns; sycamore, ash and elm were used for starting the fires in limekilns that later burned with coal; above all, charcoal, best of all oak charcoal, was used for smelting iron. Perhaps some woods *were* lost due to these activities, but the many surviving oakwoods of the western sea lochs of Argyll are testimony not to the profligate nature of the ironmasters and the landlords but to their practice of sustainability by coppice and rotational

---

* An earlier version of this paper appeared in *Scottish Forestry*, 48 (1994), pp. 244–52. Thanks to the editor for permission to reuse.

cutting (see Chapter 5). The same woods, and those in Dumbartonshire and Perthshire, provided the tannin, from their bark, for curing leather. When coal replaced charcoal for iron smelting, and chemicals replaced bark in the tanneries, then (and only then) did these woods fall into their modern disuse and decay.

So it was with timber itself, used, for example, to provide fencing materials for the farmer. The thousands upon thousands of hazel rods referred to in medieval ecclesiastical documents, and even illustrated on a fifteenth-century seal of the Douglas family, as composing a wattle deer fence, became replaced in the fullness of time by barbed wire; and sycamore, hazel or osier fencing posts became replaced by imported softwood from Scandinavia. But the Highlands are still full of formerly coppiced hazel and alder, and the Lowlands have sycamore, ash and elm on almost every farm, once used for these materials. Birch in Highland and Lowland alike had ubiquitous use; the 'Hielandman's tree' of tradition, for its soft wood could readily be cut by poor quality edge-tools, and used to make farm implements such as harrows and parts of ploughs, or turned into 'cabers' to make rafters or the crucks of stone, turf-built or wattle-walled houses. Until recently called disparagingly the 'weed of the forest' by commercial foresters, before the age of cheap metal and plastic it was the staff of life in many communities: you might not need much of it compared to the potential supply, but you needed some, and its mosaic distribution on poor land reflects this fact. Holly branches were cut in some localities for animal forage, and holly wood was useful for turnery work. Pine was also useful for cabers and planking and was a source of light from the practice of digging into its resinous core for making 'candle fir'. It had probably retreated roughly to its present very limited distribution on poor, often gravelly, soils long before 1500. Where it was accessible, it was cut and cut again. It was difficult to find trees of greater than a foot-and-a-half diameter in Rothiemurchus in the eighteenth century, but the people who were cutting them down (and letting them grow again from fresh seed) were not outsiders but local craftsmen and timber merchants floating the logs down the Spey or dragging them over mountain roads.[1]

Forget the wild wood. Trees were business and a resource so long as they were useful as timber. And if they were not useful as timber, they were emphatically useful as pasture. Again and again we hear of the

---

[1] For a survey of wood use, see T. C. Smout, Alan R. MacDonald and Fiona Watson, *A History of the Native Woodlands of Scotland, 1500–1920* (Edinburgh, 2005), esp. ch. 4.

woods being considered as invaluable winter shelter, the bottom end of a transhumance system that usually left them alone in summer; an extra two months out of doors was what a good sheltering wood meant to most seventeenth- and eighteenth-century Highland farmers, and bitter was the grumbling when the laird's foresters banished the goats, cattle, sheep and horses from an oakwood or a pine forest in the interests of timber production, when the market for wood products such as charcoal or pine planking temporarily encouraged a switch of land use. Useful, utilitarian, subject to the market – that was the predominant view of trees and the ground they grew upon.[2]

It was never, however, the only view, which returns us to the mainstream of this chapter. Trees from remotest antiquity were regarded as having spiritual attributes: Roman authorities referred generally (not specifically to Scotland) to the groves of oak associated with the Celtic druid priesthood and with acts of worship, including human sacrifice. At Jaegerspris Nordskov in Denmark, a small number of oaks thought by some to be as much as 2,000 years old could just perhaps be the relics of such a grove. In the hazy world of folklore (hazy to me because I am never quite certain of the antiquity of belief, and its lack of contamination from Victorian fantasy), every Scottish tree had its properties. Ash was the tree of life, with power to protect against charms and enchantment; rowan was another defence against evil, and its general presence outside every croft house testifies to this; apple and birch were associated with birth; elder and hawthorn with the spirit of the dead; alder with rebirth; hazel with wisdom, and so forth. Tess Darwin tells us that hazelnuts were eaten by druids seeking prophetic powers (like mescalin) and says 'the tree may have been, along with rowan, more important in Scottish druidic rites than oak'.[3] Perhaps indeed, but we know nothing whatever directly or indirectly about the rites of the prehistoric priesthood in Scotland.

All, however, is plainly not fantasy. Woods were associated in Gaelic poetry with prosperity and goodness, and in ancient Irish Brehon law the individual species of trees each had a place in a hierarchy ranging from oak, hazel, holly and yew among the nobles to rowan, birch and elm among the commoners, and blackthorn, elder and aspen among the lower orders. That Gaelic notions were similar even as late as the seventeenth century is expressed by Sìleas na Ceapaich's poem in praise of her dead lover:

[2] Ibid. ch. 5.
[3] T. Darwin, 'Sacred trees in Scottish folklore (part I)', *Reforesting Scotland*, 10 (1994), pp. 8–10.

The yew above every wood
the oak, steadfast and strong,
you were the holly, the blackthorn,
the apple rough-barked in bloom;
you had no twig of the aspen,
the alder made no claim on you,
there was none of the lime-tree in you,
you were the darling of lovely dames.[4]

Ash was undeniably special in several traditions, held in Norse myth to be Yggdrasil, the magic tree of life that touches heaven and hell. When the Earls of Orkney built Rosslyn chapel in the fifteenth century, the famous 'prentice pillar was a depiction of this tree, four dragons gnawing its roots and Christ's vine climbing up it in a spiral.

Pennant, in 1772, described the practice of midwives giving new-born babies 'sap from a green stick of ash held in the fire so that the juice oozed out onto a spoon held at the other end', which strongly confirms the tradition of it as the tree of life.[5] British forces in 1746, as part of their campaign of terror after Culloden, deliberately burnt an enormous ash, fifty-eight feet in circumference, in the churchyard of Kilmallie in Lochaber, the parish church of the Jacobite chief, Cameron of Lochiel. This tree was said to have been 'held in reverence by Lochiel and his kindred and clan, for many generations'.[6] Its destruction was as deliberate an act of cultural and ethnic vandalism as anything in Communist Romania or Taliban Afghanistan. From its girth, the Kilmallie ash was probably indeed old enough to have been a relic of pre-Christian worship.

Lochiel's ash belongs to a category of revered individual trees, which attracted attention from quite early times as natural historic sites. In the Borders, there were ancient 'march trees' that marked the boundaries between estates or communities, and others that acted as traditional trysting places. An elm at Roxburgh, called the Trysting Tree, was measured in 1796 as thirty feet in circumference. The Capon Tree, an ancient oak outside Jedburgh is still there. On the other hand, an even larger oak in the Jed Forest, 'The King of the Forest', mentioned in 1834, has long gone.[7] Perthshire had similar trees: 'The Hanged-Men's Tree' at Birnam,

[4] Translated by Derick Thomson in R. Crawford and M. Imlah (eds), *The New Penguin Book of Scottish Verse* (London, 2000), p. 221.
[5] Darwin, 'Sacred trees'.
[6] J. Walker, *Essays on Natural History and Rural Economy* (Edinburgh, 1808), pp. 16–17.
[7] Walker, *Essays*, p. 19; W. Gilpin, *Remarks on Forest Scenery and other Woodland Views*, ed. T. Dick Lauder (Edinburgh, 1834), vol. 1, p. 252.

the 'Law Tree', the trunk of an old thorn at Port of Menteith that served as the burgh cross, and 'Malloch's Oak' at Strathallan where at some time a meal merchant had (according to the local story already 200 years old in 1883) been strung up in a famine as an example to others tempted by profiteering.[8] Most famous of all was 'Wallace's Oak' in the Torwood in the parish of Larbert outside Stirling, the traditional hiding place of the Scottish hero when fugitive from the troops of Edward I. In 1723, it was described as thirty-six feet in circumference, still bearing leaves and acorns, and 'ever excepted from cutting when the wood is sold'. John Walker measured it as twenty-two feet at four feet above the ground in 1771, and observed that 'whatever may be its age, it certainly has in its ruins, the appearance of greater antiquity than what I have ever observed in any tree in Scotland ... it has been immemorially held in veneration and is still viewed in that light'.[9] By 1812, however, it was reported that 'of the famous Scots oak in the Torwood near Stirling, generally called Wallace's Oak, no trace now remains'.[10] The association with William Wallace may well have been added to a tree already old and revered for another reason. There was another Wallace's oak at Elderslie near Kilmarnock where, according to tradition, Wallace hid with three hundred men among its gigantic boughs. There were once various other Wallace trees, including a Wallace's Yew at Elcho near Perth, said in Victorian times to have been planted by the hero. In a somewhat similar fashion, the name of Pontius Pilate was once associated with Fortingall in Perthshire and its great yew, less impressive now than when it was measured by Judge Barrington prior to 1770 as fifty-two feet in circumference.[11] Thomas Hunter claimed this as 'the most aged vegetable growth in Europe, and may probably have put forth its infant shoots as far back as the time of King Solomon'.[12] The accuracy of the guess is irrelevant: it is the imaginative power of the tree that counts.

There are many eighteenth-century measurements of large trees. It does appear that it was first at this time that tree measurement became a fashion, even an obsession sometimes, as on one occasion in Angus, carried out before the magistrates. However, respect and admiration for individual trees, for whatever reason, was much more ancient than mere measurement. In a

---

[8] T. Hunter, *Woods, Forests and Estates of Perthshire* (Perth, 1883), pp. 76–7, 293, 321.

[9] A. Mitchell (ed.), *Geographical Collections Relating to Scotland made by Walter Macfarlane* (Scottish History Society, Edinburgh, 1906), vol. 1, p. 333; Walker, *Essays*, pp. 7–9.

[10] W. Nicol, *Planters Kalendar* (Edinburgh, 1812), pp. 62–3.

[11] Walker, *Essays*, p. 46; Hunter, *Woods*, p. 508; J. S. Strutt, *Sylva Brittanica or Portraits of Forest Trees* (London, 1822).

[12] Hunter, *Woods*, p. 437.

similar way, the eighteenth century saw the systematisation of admiring the woods, though it must be emphasised that in Scotland admiration for and a positive tone about woods was also entirely traditional in older writers even outside the Gaelic tradition. This point is worth labouring because it appears to contrast so markedly with what Keith Thomas tells us about Tudor and Stuart England, where the connotation of forests and wood are of gloom, darkness and danger.[13] Even for England, Thomas has probably overstated the case and it has no application in Scotland. For example, Bishop Lesley's *History* of 1578 speaks of Easter Ross 'mervellous delectable in fair forrests, in thik wodis'.[14] The topographical accounts collected by Macfarlane include Timothy Pont's descriptions from the 1590s of Kintail as a 'fair and sweet countrey watered with divers rivers covered with strait glenish woods', and of Loch Maree as 'compassed about with many fair and tall woods as any in all the west of Scotland . . . All thir bounds is compas'd and hem'd in with many hills but thois beautifull to look on, thair skirts being all adorned with wood even to the brink of the loch.'[15] Robert Gordon of Straloch, half a century later, similarly describes a hill near Ballater in upper Deeside:

> its rocks and its summits to the very highest point occupied by a beautiful forest of tall evergreen firs of immense size, while the pleasing greenery of the limes and birches clothes the slopes of the mountains and the plains nearest the river . . . among the numerous forests with which the river is wooded, particularly in the upper parts, this mountain is very pleasant to see.[16]

This is a translation from the Latin, and the limes, *tilia*, are unlikely: but probably Gordon could not find a Latin word for rowans. This love of trees and woods encouraged the nobility at least from the later seventeenth century to plant extensively, though none more oddly than the Earl of Wigton who laid out his woods in blocks to represent the disposition of the troops on one of Marlborough's battlefields.

What the eighteenth century taught, however, was to look upon trees as 'picturesque', in the technical sense – that is to say, as components in a composed landscape-picture. Thus the guru of the picturesque, William Gilpin (1794), wrote:

[13] K. Thomas, *Man and the Natural World: Changing Attitudes in England, 1500–1800* (Harmondsworth, 1984), pp. 194–7.
[14] P. Hume Brown, *Scotland before 1800 from Contemporary Documents* (Edinburgh, 1893), p. 142.
[15] Mitchell (ed.), *Macfarlane's Collections*, vol. 2, p. 283.
[16] Ibid. p. 300.

> Remark the form, the foliage of each tree
> And what its leading feature. View the oak,
> Its massy limbs, its majesty of shade;
> The pendent birch, the beech of many a stem,
> The lighter ash, and all their changeful hues
> In spring or autumn, russet, green or grey.[17]

Sometimes this approach led to an intense scrutiny of detail, as in Gilpin's account of moss on an oak:

> we may observe also touches of red; and sometimes, but rarely, a bright yellow, which is like a gleam of sunshine; and in many trees you will see one species growing upon another; the knotted brimstone-coloured fringe clinging to a lighter species; or the black softening into red.

Usually, in spite of its exactness, it reduced nature to ornament. This is Gilpin on rowans:

> There, on some rocky mountain, covered with dark pines and waving birch, which cast a solemn gloom over the lake below, a few mountain ashes joining in a clump, and mixing with them, have a fine effect. In summer the light green tint of their foliage, and in autumn the glowing berries which hang clustering upon them, contrast beautifully with the deeper green of the pines: and if they are happily blended, and not in too large a proportion, they add some of the most picturesque furniture with which the sides of those rugged mountains are invested.[18]

And often it was eccentrically pompous, as in Gilpin's view of beech trees: 'the branches are fantastically wreathed and disproportioned, twining awkwardly among each other . . . in full leaf it is equally unpleasing: it has the appearance of an overgrown bush'. Gilpin's editor in 1834, Sir Thomas Dick Lauder, was dismissive of that particular opinion of his author: 'a noble beech', he retorted, is 'one of the most magnificent objects of God's creation'.[19] In truth, for the eighteenth-century Scottish Improvers, the beech was a hallmark of their activities in the countryside,

---

[17] W. Gilpin, *Essays on Picturesque Beauty* (London, 1794), p. 100.
[18] Gilpin, *Remarks*, pp. 51, 89.
[19] Ibid. pp. 98–100.

almost an icon. It was modern, introduced and English, itself a recommendation; it was impressive and orderly, crowding out the redundant undergrowth much as they crowded out redundant tenantry; it was useful, making excellent bobbin wood for rural industry; and it was generally held to be just as beautiful as Sir Thomas Dick Lauder thought. An overgrown bush, indeed!

It is no wonder that Wordsworth and his friends, who considered themselves to have a better appreciation of wild nature in the raw, for the sublime and the terrific rather than the picturesque, were very much more dismissive than Dick Lauder of the preciousness of William Gilpin and his school. The romantics were also watchful of the wanton destruction – as they regarded it – of old timber. The impecunious Duke of Queensberry fell foul of their poetic wrath. In 1796, Burns, surveying 'the destruction of the woods near Drumlanrig', made the sprite of Nith exclaim, 'the worm that gnaw'd my bonny trees, That reptile wears a ducal crown'. And Wordsworth in 1803, encountering similar destruction of 'a noble horde. A brotherhood of venerable trees' near Neidpath in Peeblesshire, commenced his denunciation with the words: 'Degenerate Douglas! oh, the unworthy Lord!' Such protests, on purely aesthetic grounds, about what was after all only the commercial felling of mature timber, seem quite new, but they were no passing fashion. Sir Thomas Dick Lauder, for example, spoke with horror about the devastation of the ancient Caledonian pines at Glenmore and Rothiemurchus following the high prices for Scottish deal planking during the Napoleonic Wars, though he was enough of a forester to comment also on the abundant natural regeneration that was following.[20] The Wordsworthian attitude lies behind today's Tree Preservation Orders that treat selected trees exactly like scheduled historic buildings, as having an overriding public value for their aesthetic and historic associations. Though TPOs do much good, unlike buildings but like ourselves, the mortality of trees is ultimately impervious to bureaucratic command.

By the early nineteenth century it was a very well-established idea that trees and woods, and walks among them, had a value as refreshment to the spirit, apart from and additional to any commercial value. This did not just apply to ornamental plantations, but also to natural woods. The point was neatly underscored by a dispute in the Grant of Rothiemurchus family in the early nineteenth century, when the laird sought to vary the terms of an entail that forbade him to cut the woods for twenty years (basically imposed by his father to give the forest time to recuperate from earlier exploitation). His relatives, while not opposing him in general, sought to continue protection

---

[20] Ibid. pp. 175–6.

for the pines around Loch an Eilean and the other waters, because they were 'part of the pleasure grounds of the manour of the Doune', and therefore an amenity that the whole family had a right to enjoy and an interest in preserving.[21]

Of course, the idea of the woods as a locus for recreation has grown and grown, culminating in the twentieth century in the creation by the Forestry Commission of a series of Forest Parks. This commenced with Ardgarten and what became the Argyll Forest Park in 1936, adding to it Glentrool and Glenmore (and three in England) by 1947, and the Queen Elizabeth Forest Park in the Trossachs in 1951 and the Border Forest Park in 1955. At first, these mostly utilised the unplantable upper land, but it was quickly realised that the forests themselves, planted Sitka, lodgepole pine and Norway spruce though they may be, had popular appeal: indeed Lord Robinson, chairman from 1932–52, expressed incredulity when he read the Dower Report in 1944, for, he said, everyone preferred walking through forests to walking through open land. Earlier, he had fired the prime minister, Ramsay Macdonald, with enthusiasm for his plans for public recreation in the 1930s, to the irritation of the Treasury.[22] But Robinson's enthusiasm was politically shrewd, and the Treasury's concern understandable. Forestry, as an activity heavily subsidised by the taxpayer, needed political allies and a high public profile: where better to seek both than in the democracy of ramblers and families on Sunday outings? It may be not too much to say that when in the 1990s the threat came of privatisation, it was above all the value of Forestry Commission woods for popular recreation that saved public-sector forestry from complete extinction.

Let us return to the Wordsworthian age. An influential variety of romantic aesthetic theory was that associated with Archibald Alison's work of 1790, and re-expressed by Thomas Dick Lauder in an essay of 1834. It held that 'the mere forms and colours' that compose the visible appearance of nature 'are no more capable of exciting any emotion in the mind, than the form and colour of a Turkey carpet. It is sympathy with the present or the past, or the imaginary inhabitants of such a region, that alone gives it either interest or beauty.' Dick Lauder went on to apply this to Highland scenery:

Then there is the impression of the Mighty Power, which piled the massive cliffs upon each other, and rent the mountains asunder, and

[21] T. C. Smout and R. A. Lambert (eds), *Rothiemurchus: Nature and People on a Highland Estate, 1500–2000* (Dalkeith, 1999), p. 66.

[22] National Archives of Scotland: Forestry Commission 9/1, minutes 19 Jan. 1944; J. Sheail, *Rural Conservation in Inter-war Britain* (Oxford, 1981), pp. 172–86.

scattered their giant fragments at their base; and all the images connected with the monuments of ancient magnificence and extinguished hostility – the feuds, the combats, and the triumphs of its wild and primitive inhabitants, contrasted with the stillness and desolation of the scenes where they lie interred – and the romantic ideas attached to their ancient traditions, and the peculiarities of their present life – their wild and enthusiastic poetry – their gloomy superstitions – their attachment to their chiefs – the dangers and the hardships, and enjoyments of their lonely huntings and fishings – their pastoral shielings on the mountains in summer . . .[23]

And so on.

At first sight this is some distance from the American wilderness school associated with Emerson, Thoreau and Muir, where admiration for nature increases in proportion to its quality of being 'untouched by human hand', though Alison and Emerson could presumably have equally enjoyed 'the impression of the Mighty Power, which piled the massive cliffs upon each other'. But the Alison view, itself surely rooted in Adam Smith's *Theory of Moral Sentiments* as well as in Edmund Burke's aesthetic ideas, has always appealed to the Scottish eye. Though his language is different, it appeals to the modern appreciation of cultural landscapes: a landscape is its people and its history, and not the patterns of a Turkey carpet. In the Alison view, all landscapes are historic landscapes.

The idea of a Great Wood of Caledon shows how in practice landscape as history and landscape as wilderness can actually reinforce each other and become politically potent. We owe the notion, of course, originally to a rough map by Ptolemy in the second century AD, and to a few chance remarks of Roman writers about woods in that part of the island they called Caledonia, which itself was a place not always completely fixed in their minds. David Breeze has shown how vague and ambiguous were the observations of Tacitus and other writers, and how the archaeological record quite fails to support the popular view of Roman Scotland covered with wall-to-wall forest.[24] We can readily accept McVean and Ratcliffe's account of the original plant communities and their confirmation that most of the land was covered with forest 5,000 years ago, hardwoods broadly predominating in the south and west and pine in the north and east. But the precise ways in which the forest was removed, either by man or by natural

---

[23] Dick Lauder's prefatory essay to Gilpin, *Remarks*, pp. 11–12.
[24] D. J. Breeze, 'The great myth of Caledon', *Scottish Forestry*, 46 (1992), pp. 331–5, reprinted in T. C. Smout (ed.), *Scottish Woodland History* (Edinburgh, 1997), pp. 47–50.

conditions such as deteriorating climate and peat formation, or by a combination of the human and the natural, are not known in detail – although palynological and archaeological investigation have shown convincingly that much of the destruction took place before Roman arrival, and that it involved both climate change and human agency. In the Lowlands at least, deforestation was very extensive. In the Highlands, more wood would no doubt have remained, but by no means the unbroken, trackless forest of common belief: in Breeze's view only 'isolated pockets of the original pine and birch woods' survived to Roman times. This may be an overstatement, but only more extensive palynological investigation can establish a proper chronology.[25]

But that is not the point here. A tale has come down to the late twentieth century that there was once a great wood, especially a great Highland wood of pine, that protected the Picts from the Romans; that its destruction was the work of aliens and of greedy landowners; that it had been ecologically benign and could be so again; and that it would be a worthy work for the twenty-first century to right the wrongs of earlier centuries and to recreate such a wood. It is worth tracing how this developed. Never mind the actuality, feel the myth.

I have described the rise and elaboration of this tale at length elsewhere, so here it is enough to abbreviate it.[26] It appears in modern form in Hector Boece, whose *History of Scotland* was published in 1527 and who imaginatively described the great Roman wood as having once stretched north from Stirling as far as Lochaber, full of white bulls with 'crispand mane, like feirs lionis'.[27] This became repeated throughout the next three centuries, though without particular emphasis and always as something in the past.

It got a new lease of life in two stages. Towards the end of the eighteenth century, the revival of interest in Roman antiquities encouraged scholars to regard the fossil oaks and pines that were excavated from time to time in Lowland bogs as well as in Highland ones, as relic proof of the Great Forest of Caledon. Thus John Walker in 1808 remarked that 'the Caledonian Forest, mentioned by Tacitus, seems to have extended

---

[25] D. N. McVean and D. A. Ratcliffe, *Plant Communities of the Scottish Highlands* (London, 1962); R. Tipping, 'The form and fate of Scottish woodlands', *Proceedings of the Society of Antiquaries of Scotland*, 124 (1994), pp. 1–54; J. H. Dickson, 'Scottish woodlands: their ancient past and precarious future', *Scottish Forestry*, 47 (1993), pp. 73–8; D. J. Breeze, *Roman Scotland* (London, 1996), p. 97.

[26] T. C. Smout, *Nature Contested: Environmental History in Scotland and Northern England since 1600* (Edinburgh, 2000), ch. 2.

[27] Hume Brown, *Scotland before 1800*, pp. 80–1.

many miles above and below Stirling'.[28] We now know from carbon dating that most of these fossils are about 4,000 years old. They date from the prehistoric woodland maximum, before the coming of farming, and appear to have fallen victim to climate change millennia before the Romans came.

The second puff for the Great Wood came from Victorian romantics, especially John and Charles Allan, the so-called 'Sobieski Stuarts', who with much elaboration of detail caught the public imagination in 1848 with talk of the *Caledonia Silva* as 'the great primeval cloud, which covered the hills and plains of Scotland before they were cleared'.[29] In no time at all this fantasy became established as fact. Nineteenth-century historians said it must have been like the African jungle, and invited their readers to imagine the Picts fighting the Romans amid its fastnesses. Twentieth-century ecologists took up the tale, none more influentially than Frank Fraser Darling, who in 1947 partly blamed the ravaging Vikings, but mainly the seventeenth- and eighteenth-century ironmasters, sheep farmers and timber speculators. He said, 'the imagination of a naturalist can conjure up a picture of what the great forest was like: the present writer is inclined to look upon it as his idea of heaven'.[30] Its destruction, he was to suggest, left the Highlands as a degraded landscape.

Later commentators elaborated the message in book and film as a polemical argument.[31] Until capitalism and the English came (at some point after 1600), the Great Wood of Caledon had covered the Highlands. A great sin against nature had been committed in destroying it, and it would be a worthy reparation to put it back again by planting native woods, especially the Caledonian pine, which became by association almost synonymous with the Caledonian forest. The Green Party espoused this, the journal *Reforesting Scotland* and the charity Trees for Life, among others. Objectively considered, most of this is actually nonsense. Yes, most of Scotland probably was covered with wood 5,000 years ago. There is no evidence that it remained intact when the Roman's came, 2,000 years ago, still less that it stayed substantially pristine in the Highlands until the seventeenth century; even Hector Boece had put it in the past.

It is frequently said that only 1 per cent of the original Caledonian Forest

[28] J. Walker, *An Economical History of the Hebrides and Highlands of Scotland* (Edinburgh, 1808), vol. 2, p. 188.

[29] J. S. Stuart and C. E. Stuart, *Lays of the Deer Forest* (Edinburgh, 1848), vol. 2, pp. 256–7.

[30] F. Fraser Darling, *Natural History in the Highlands and Islands* (London, 1947), pp. 57–8.

[31] For example, H. Miles and B. Jackman, *The Great Wood of Caledon* (Lanark, 1991) and the associated film.

is now left, or alternatively, that only 1 per cent of Scotland still has its original woodland cover. Why should what the Scottish countryside looked like five millennia ago in a different climatic regime and only a sparse human population of hunter gatherers be relevant to what it ought to look like now? Nevertheless, the image of restoring the Caledonian Forest proved a potent one for politicians and conservation charities in the late twentieth century, though it is less powerful at the moment than it was a decade ago. The Heritage Lottery Fund, for example, parted with millions to put back as much of Caledonia Silva as they could in time for the year 2000. I do not think that planting native woodland is bad, though there is a question of how much to plant and where to plant it. But I do think using bad history is bad, and leads to bad results when you begin to consider how much and where. Unfortunately, history does not have to be accurate to be influential: too much of human experience suggests that the more inaccurate it is, the greater its leverage.

The Great Wood of Caledon is a big historic landscape of the mind. But the human imagination is also stirred by little landscapes, in the shape of old trees. We are all moved by forest giants, and we want to ask the questions, how big are they, how did they get there, what have they witnessed beneath their shades? The first of these questions began to be asked systematically in the eighteenth century, when measuring these trees became something of a gentlemanly pastime and has remained an interest for many people ever since. The second, their provenance, could only be authentically answered when foresters began to keep good records, very seldom before around 1830: for older trees there might be, or come to be, a 'tradition'. The third, their experience, was unanswerable except in 'tradition', and a few such traditions are manifestly old, like the tale of Wallace's oaks and Malloch's oak. Most, however, appear to be like the full-blown Caledonia Silva, Victorian inventions.

The Forestry Commission, in association with the Tree Council, recently published a lovely book on 'heritage trees' and it has been widely distributed by the Woodland Trust and others anxious to preserve such splendid things, not least for their value for the wildlife that makes use of them.[32] The authors list 130 remarkable trees, with illustrations and informative text. About half are relatively modern, mainly North American conifers grown from seed in the eighteenth or more often the nineteenth century, such as the original larches at Dunkeld, the original Douglas firs at Scone and Drumlanrig, the Ardkinglas grand fir and the Drumtochty Sitka spruce: all awe-inspiring giants with well-attested histories.

---

[32] D. Rodger, J. Stokes and J. Ogilvie, *Heritage Trees of Scotland* (Edinburgh, 2006).

The other half are mainly older, and among their boughs, 'traditions' run riot. There are, for example, trees said to have been planted to mark the death of James II (Roxburgh, holly), the Battle of Flodden (Coldstream, sweet chestnut), Cromwell's march on Perth (Bridge of Earn, sweet chestnut) and the Union of 1707 (North Berwick, beech). Others are said to have been planted by famous people – Rizzio (Melville Castle, sweet chestnut), Mary Queen of Scots in 1561 (Cumbernauld, sweet chestnut), 1563 (St Andrews, thorn) and 1565 (Balmerino, sweet chestnut), and James VI (Scone, oak and sycamore). Mary is said to have danced round a thorn (Pitcaple, Aberdeenshire). John Knox is said to have preached beneath two yews (Ormiston and Langbank). David I is said to have planted a deer park at Cadzow and there are other thirteenth-century attributions. Famous events are supposed to have happened in their shade: Mary nursing Darnley back to health (Glasgow, Crookston, a sycamore centuries too young for this to have happened), the murder of Darnley plotted (Whittingehame, yew), the Treaty of Union of 1603 drafted (Loudon, yew – there was no such treaty). Rob Roy and Bonnie Prince Charlie are both said to have hidden in Eppie Callum's oak (Crieff – there is no evidence that they ever came near Crieff to hide from anyone). There are 'Dule trees', said to be gallows (at Blairquhan, Ayrshire and Huntley, Aberdeenshire): but the original meaning of 'dule' was boundary, and the first recorded use of the term as a gallows is given by the *Dictionary of the Older Scottish Tongue* as 1864.

There is, in fact, not a shred of written evidence that any of the above associations are true, and the authors of the *Heritage Trees* wisely attribute them all to tradition, though occasionally falling for the temptation to muse on how thrilling it is to have such a link with Scotland's past. Most of the associations are not in the least bit likely. The first tree I know of to have been planted by a notable person was Benjamin Franklin's planting of an American tree in Paris botanic gardens late in the eighteenth century, and I can think of no record of trees being planted to mark a notable event before the nineteenth century. Nor is it believed that Knox ever preached out of doors, though there are plenty of records of him preaching in churches.

There is tradition and there is invention, and there is the invention of tradition. The stories listed above do not appear in the *Statistical Account of Scotland* of the 1790s (which asked for information on antiquities), or in the early accounts of notable trees (though other traditions did, such as Wallace's oak at Elderslie and the Fortingall yew's link with Pontius Pilate). Most of them are probably nineteenth-century invention: another fruit of the Romantic imagination. A few may be better based: people at least did dance round thorns, and some got hanged on trees, though not necessarily boundary trees. This all seems to confirm what is evident from the story of the Great Wood of Caledon: Victorians loved their imaginations to run wild. And it confirms what we know about ourselves: we are suckers for a good tale.

Fig. 10 A little owl in Co. Durham, July 2008. Introduced into England late in the nineteenth century, it was for many years regarded as vermin by farmers and gamekeepers. After careful (and pioneering) studies of its diet, it was concluded that it mostly ate insects, worms and mice, and was placed on the schedule of protected species in 1954. Courtesy of David Brown Website Design, Imaging and Photography: www.daveb.co.uk.

CHAPTER 10

The Alien Species in Twentieth-Century Britain: Inventing a New Vermin*

Much recent conservation management in the United Kingdom (and indeed throughout the world) is concerned, implicitly or explicitly, with the distinction between a native and a non-native species. The most generally accepted definition in Britain of an alien or non-native species is one introduced by human action since the last ice age, 10,000 years ago.[1] To the lay person, and often to the scientist, this can result in some unexpected classification. The familiar sycamore and sweet chestnut, the rabbit and the black rat are all alien species, though they have been in the UK for between 500 and 2,000 years; so is the brown rat and *Rhododendron ponticum* (from the eighteenth century), a whole raft of North American conifers (from the nineteenth century) and, less unexpectedly, the American mink, the ruddy duck and the muntjak deer (from the twentieth century). On the other hand, the capercaillie, the goshawk, the white-tailed eagle and the European beaver are native species, though they all became extinct in Britain between the sixteenth and the twentieth century and owe their reintroduction to human intervention. The botanist is especially at a loss because of the real difficulty of distinguishing between what might have been here before Neolithic people arrived with the paraphernalia of farming, and what came with them, or subsequently in the Bronze Age, the Iron Age and with medieval monks and travellers. The term 'archaeophytes' has been devised to cover those species which it is suspected are not native but have been here for 500 years.[2]

---

* This was part of a symposium on alien species in historical context held at the European University, Florence, and published in *Landscape Research*, 28 (2003), pp. 11–20. Thanks to Carfax Publishing for permission to reproduce.
[1] M. B. Usher, 'The nativeness and non-nativeness of species', *Watsonia*, 23 (2000), pp. 323–6; D. A. Webb, 'What are the criteria for presuming native status?', *Watsonia*, 15 (1985), pp. 231–6.
[2] P. Macpherson, J. H. Dickson, D. G. Kent and C. Stace, 'Plant status nomenclature', *BSBI News*, 72 (1996), pp. 13–16; M. Schever-Lorenzen, A. Elend, S. Nöllert and E.-D. Schulze, 'Plant invasions in Germany', in H. A. Mooney and R. J. Hobbs (eds), *Invasive Species in a Changing World* (Washington, DC, 2000), pp. 351–68.

Alien species are generally denied statutory protection, usually regarded by conservationists with deep suspicion and certainly as second-class citizens, and often (and often justifiably) as a most serious threat to the natural heritage. Internationally, the impact of alien species leads to the homogenisation of the earth's biotic communities and is regarded by the International Union for the Conservation of Nature (IUCN) as one of the two most serious threats to the world's biodiversity, the other being habitat loss and fragmentation.[3]

This perspective is a comparatively recent one. In the later nineteenth century and the early part of the twentieth century, there was not necessarily any adverse connotation about being an introduced or alien species. Animals like American beaver or red-necked wallabies, and insects like the map butterfly were still being introduced by naturalists into the wild, to add interest and variety to the native fauna, often unsuccessfully but usually without misgivings. Game birds were still being introduced for sport, as they had been for centuries. Plants were being introduced for ornament or for utility, like the Japanese laurel for the garden and the Japanese larch for the forestry plantation.[4] The grass *Spartina townsendii* was a polyploid hybrid species that arose in Southampton Water around 1870 from a cross between the native *S. maritima* and the American *S. alternifolia*. It proved a much more aggressive coloniser of tidal mudflats than either of its parents, and partly on its own accord and partly due to human intervention spread from Southampton to estuaries around the British Isles, modifying to a degree their ecosystem.[5] Around 1937, the Linnean Society of London and the Royal Horticultural Society of Dublin protested against leaflets exhorting lovers of nature to 'beautify the countryside' and offering seeds of alien plants for the purpose. The problem, as seen on that occasion by the distinguished Irish naturalist Robert Lloyd Praeger, was not so much the risk of introducing new weed species, but of confusing the natural patterns of plant distribution, which since the days of Wallace and Darwin had been so fruitful an object of scientific investigation.[6]

However, most alien species were introduced by accident, as stowaways on boats, and in ever-accelerating numbers over the last three centuries as international trade spread. They were sometimes a menace to agricultural crops – for example, the Colorado beetle that threatened the potato – or a menace to

[3] J. A. McNeely, 'The future of alien invasive species; changing social views', in Mooney and Hobbs (eds), *Invasive Species*, pp. 171–98; M. Williamson, *Biological Invasions* (London, 1996).

[4] S. Tachibana and C. Watkins, 'The assimilation, naturalisation and hybridisation of Japanese trees and shrubs in Britain 1700–1920', in T. Mizoguchi (ed.), *The Environmental Histories of Europe and Japan* (Nagoya University, 2008), pp. 185–91.

[5] C. S. Elton, *The Ecology of Invasions by Animals and Plants* [new edition with foreword by D. Simberloff] (Chicago, 2000), p. 26.

[6] R. L. Praeger, *The Way That I Went* (Dublin, 1980), pp. 363–4.

health, like *Pasturella pestis*, the bubonic plague bacillus suspected of being the cause of the medieval Black Death that last reached Glasgow from the Far East in 1900 and killed fifteen people. Some have arrived by strange accident: the freshwater flatworm *Phagocata woodworthi* arrived in Loch Ness from North America apparently on the equipment used by monster hunters; it is regarded as a considerable potential threat to native tricladid fauna.[7] But, welcome or not, alien species came, and still come, in enormous numbers. To take plants, in 1994 it was reckoned that 3,586 non-native species had been recorded beyond the garden in the British Isles, of which 885 had currently established naturalised populations. Of Britain's mammals, twenty-one out of forty-nine established species are aliens. The situation is similar abroad: the USA has about 4,500 exotic species in a free-ranging state, of which 20 per cent are considered economically or ecologically harmful.[8] And the ecological damage done to biodiversity by alien species is indeed great – nowhere more than on remote islands such as Hawaii or Tristan da Cunha and, continentally, among freshwater fish with isolated populations. In the USA, 63 per cent of freshwater fish are under threat from competition by alien species, and 23 per cent threatened by hybridisation of species or sub-species. Eclipsing all is the expected extermination of about 200 of the 300 species of cichlid fish endemic to Lake Victoria, through the introduction in 1959 of Nile Perch as a sports fish: 'never before has man in a single ill advised step placed so many vertebrate species simultaneously at risk of extinction and also, in doing so, threatened a food resource and traditional way of life of riparian dwellers'.[9]

Nevertheless, the damage varies between groups. Terrestrial plant introductions, at least in Europe, have not proved a problem to native species on a scale enough to threaten extinction.[10] Budiansky even claims – in relation to all biota – that 'no mainland species has ever been extirpated solely as a result of the introduction of an alien species',[11] though that does not perhaps take account of fish.

---

[7] P. S. Maitland and C. E. Adams, 'Introduced freshwater invertebrates in Scotland: enhanced biodiversity or a threat to native species', *Glasgow Naturalist*, 23 supplement (2001), pp. 26–34.

[8] Usher, 'Nativeness'; E. J. Clement and M. C. Foster, *Alien Plants of the British Isles* (London, 1984); McNeely, 'Alien invasive species', pp. 173, 176.

[9] E. O. Wilson, *The Diversity of Life* (Cambridge MA, 1992), p. 256.

[10] J. H. Dickson, 'Alien vascular plants in Scotland', *Glasgow Naturalist*, 23, supplement (2001), pp. 2–12; K. R. Edwards, 'A critique of the general approach to invasive plant species', in U. Starfinger, K. Edwards, I. Kowarik and M. Williamson (eds), *Plant Invasions: Ecological Mechanisms and Human Responses* (Leiden, 1998).

[11] S. Budiansky, *Nature's Keepers: the New Science of Nature Management (New York, 1995)*, p. 129.

At the start of the twentieth century, however, the overriding category was not whether a species was alien or not, but whether it was 'vermin' or not. Vermin were species that for one reason or another were regarded as undesirable, usually because they were considered pests of agriculture or horticulture, or enemies to game preservation. Vermin in Great Britain included virtually everything with a hooked bill, mammals such as grey and common seals, otters and badgers as well as foxes and moles, and small birds with a liking for buds and fruit, like bullfinches and blackbirds, sometimes even blue tits. Some vermin were aliens, like the rabbit, the black rat and the brown rat. That they were foreigners was sometimes part of the indictment against them, but native vermin and alien vermin were treated the same. Red squirrels as well as grey squirrels were slaughtered in their thousands in defence of forestry.

When James Ritchie in 1920 wrote his deeply original book *The Influence of Man on Animal Life in Scotland*, in a far-sighted passage he warned against the 'many thoughtless introductions', and pointed 'a warning finger at the naturalist and reformer who, by introducing animals would revise Nature's order, and by short cuts and unimaginative experiments tends to make a wilderness where he had looked for a paradise'. He used the term 'undesirable aliens', but not in any systematic or defined way, and preferred the expressions 'deliberate' or 'accidental' 'introductions'. He observed correctly that 'the alien stowaways which become established in a country include more economic pests than the native fauna they invade'. But the main theme of his book was much wider: about how habitat alteration by human intervention, by domestication of animals as well as by introductions and persecutions, had irretrievably altered nature.[12]

The main agenda of British conservationists in the first half of the twentieth century, apart from seeking to establish nature reserves, was to reduce the vermin list as much as possible on the grounds that the damage done was much exaggerated. The history of this campaign has been relatively little studied, but it is interesting that one of the first and best scientific papers on the food of a raptor, W. E. Collinge's work on the little owl, was an investigation into an alien species that exonerated it from causing significant harm either to native songbirds or to game chicks.[13] Scientists often argued that to define so many species as pests was to interfere with a 'balance of nature', which would actually be counterproductive to human interests. Tansley,

---

[12] J. Ritchie, *The Influence of Man on Animal Life in Scotland* (Cambridge, 1920), quotations on pp. 300, 419.

[13] W. E. Collinge, 'Food of the little owl', *Journal of the Ministry of Agriculture*, 28 (1924–7), pp. 1022–31, 1133–40; J. Sheail, *Nature in Trust: the History of Nature Conservation in Britain* (Glasgow, 1976), p. 51.

the leading English ecologist of his day, blamed the failure of the native oakwoods to regenerate in Britain on excessive numbers of mice and voles, an imbalance that he thought had been brought about by the destruction of birds of prey and stoats and weasels by gamekeepers. This was not a good explanation, but the belief in balance was characteristic of its time.[14] Thanks as much to a rising popular interest in nature study as to the burgeoning science of ecology, the efforts of the conservationists were remarkably successful. By the 1950s, statutory protection was extended over the vast majority of birds, over many mammals and some other species. At least one of these protected species, the little owl, was a relatively recent introduction.

Campaigns against the remaining vermin continued unabated by landholders and public health officers, and often the vermin involved were indeed alien species like muskrat and coypu (both eventually exterminated), black rats and brown rats and, most determinedly and contentiously of all, rabbits – the last through the introduction of yet another alien species, the myxoma virus.[15] But other native species were equally treated as vermin, such as wood pigeons, carrion crows, magpies and moles, and still little distinction was drawn between the two classes. No one in the 1950s tried to justify biological warfare against the rabbit on the grounds that it was not truly British, or to applaud the changed appearance and diminished biodiversity of the Suffolk heaths on the grounds that they were returning to a pre-rabbit, pre-Roman England. The discussion was in terms of enhanced food production *versus* negative habitat change. Conservationists are still ambivalent about the rabbit.

Scientists writing in these decades in the tradition of Ritchie in the New Naturalist series, especially Max Nicholson (*Birds and Men*) and Edward Salisbury (in the significantly titled *Weeds and Aliens*), continued to give equal weight to native and non-native species. Nicholson emphasised how the little owl and the red-legged partridge had found a niche in British farmland ecology without harm to other species, and Salisbury emphasised that common ragwort and bracken were native species every bit as aggressive and pestiferous to agriculture as any introduction.[16]

The most important and far-sighted writing of the time, however, was Charles Elton's *The Ecology of Invasions by Animals and Plants* (1958). Elton was a highly distinguished mammal ecologist at the Bureau of Animal Population at Oxford. His book, which developed from three broadcast

---

[14] A. G. Tansley, *The British Isles and their Vegetation* (Cambridge, 1953).

[15] N. W. Moore, *The Bird of Time: the Science and Politics of Nature Conservation* (Cambridge, 1987), pp. 121–43.

[16] M. Nicholson, *Birds and Men* (London, 1951), pp. 67–70, 92–5; E. Salisbury, *Weeds and Aliens* (London, 1961), pp. 199–207.

talks on the BBC Third Programme, is recognised now as having laid the cornerstone for the ecological study of invasive species. In it, Elton identified the main characteristics of the subject, including the special problems invasions posed to island populations, the difficulty of identifying which of the many invaders would eventually pose a problem, and the risk involved in introducing as biological controls other non-native species ('counterpests') to combat existing invaders. His perspective on the problem, however, was a measured one. He believed that the best defence against damaging invasions was a robust native ecosystem. He believed that in time some invasions might moderate their destructive force, as the pondweed *Elodea canadensis* that had choked British rivers in the nineteenth century came to do in the twentieth century. And he believed that there might even be a place for the non-native species in a healthy ecosystem:

> I believe that conservation should mean the keeping or putting in the landscape of the greatest possible ecological variety – in the world, in every continent or island, and so far as is practicable in every district. And provided the native species have their place, I see no reason why the reconstitution of communities to make them rich and interesting and stable should not include a careful selection of exotic forms, especially as many of these are in any case going to arrive in due course and occupy some niche.[17]

Then, in the 1970s and 1980s, a number of things happened, operating both at the level of popular perception and of science itself, which changed the tone of the discussion. Firstly, the rapid transformation not only of bare uplands but also of existing ancient woods into plantations of Sitka spruce and lodgepole pine created a conservation backlash in favour of native broadleaves and the native Scots pine. Aesthetic distaste for non-native trees in the countryside went back at least as far as Wordsworth,[18] and added fuel to concern about the ecosystems that the new woods replaced. 'Save our landscape from alien conifers', became a popular cry. Secondly, conservationists trying to manage nature became increasingly concerned about the impact of certain species of fauna on others. In fact, they had problems *both* from native species and introduced ones, but they were rather reluctant to admit to the first. Because they still had some faith in natural balance, and in predator density being prey dependent, they were reluctant to admit that one native species could more than temporarily affect the population of

---

[17] Elton, *Ecology of Invasions*, p. 155.
[18] R. Noyes, *Wordsworth and the Art of Landscape* (New York, 1973).

another. So they were slow to take action against carrion crows and foxes on nature reserves, until compelled by overwhelming evidence in individual cases of the need to do so. Catastrophic damage by foxes to the biggest gullery in Britain, the black-headed gull colony at Ravenglass in Cumbria was one example among several.[19] Conservationists were similarly slow to admit that hen harriers could ever impact on red grouse numbers, until obliged by exhaustive research to acknowledge circumstances where the so-called 'predator trap' could come into play to prevent grouse recovery, even though this was not the prime or initiating cause of grouse decline.[20] Landowners of course were even more dismayed by this research, as they had maintained that the harriers were the prime cause of grouse decline and not the over-exploited moorland habitat for which they were themselves responsible.

However, with alien species, conservationists now found no problems in identifying an impact. American mink were having serious effects predating breeding waterfowl and the native water vole. American grey squirrels were rapidly driving out native red squirrels; hedgehogs introduced into Scottish islands, where they were not native, destroyed internationally important concentrations of breeding waders; muntjak deer wrecked the floral and insect biodiversity of Monks Wood National Nature Reserve in Cambridgeshire and many another English wood; *R. ponticum* poisoned the ground in Welsh oakwoods – and so on.[21]

From species like these, against which the conservation manager waged war because the effects were indeed serious and self-evident, the list of new vermin (although the term was never used) became gradually extended to justify intervention against many other species and sub-species of non-native origin. The sycamore, introduced in the earlier sixteenth century or before, became condemned for its invasiveness of native broadleaf woodland and accused (wrongly as it turned out) of harbouring too few native insects. The Sika deer attracted a shoot-to-kill policy not because it drove out native red deer but because it bred with them and threatened their 'genetic integrity'. Planted trees of non-local provenance but of native species – like oaks from Poland of the same species as pedunculate English oak but of a different genetic strain – stood condemned as unsuitable for

---

[19] D. Simpson, 'Whatever happened to the Ravenglass gullery?', *British Wildlife*, 12 (2001), pp. 153–62.

[20] DETR and JNCC, *Report of the UK Raptor Working Group* (London, 2000); P. Hudson, *Grouse in Space and Time: the Population Biology of a Managed Gamebird* (Fordingbridge, 1992); Budiansky, *Nature's Keepers*, pp. 208–16.

[21] P. Marren, *Nature Conservation: a Review of the Conservation of Wildlife in Britain, 1950–2001* (London, 2002).

new English forests. One new trend in conservation was an emphasis on saving not only species but 'genotypes', no doubt a consequence of the growth of molecular biology and increasing knowledge of DNA. A fuller understanding of genetic diversity within species, and indeed of the uncertain boundaries between many species, led to a call to preserve the fullest possible range of natural genetic diversity. In the UK Biodiversity Action Plan, that followed the United Nations convention in Rio de Janeiro, these principles were fully endorsed. Conservation in the UK plan was to be only for native species and genotypes. Furthermore, the definition of alien species seemed to be self-contradictory: 'A species which does not naturally occur within any area (usually a country) and which has either arrived naturally, or more usually as a result of man's intervention . . .'[22] If a species has arrived naturally does it not then occur naturally? The formulation is unhelpful and obscure.

In due course, all this led to the declaration by the Department of the Environment in 1998 of a war of extermination against the American ruddy duck. This bird was introduced into Britain in the 1940s, subsequently escaped from captivity and began to breed in the Birmingham area: in those undiscriminating days, it was thought so attractive that the West Midlands Bird Club adopted it for its logo. By 2000, the British population was about 6,000 birds. In due course, a few spread to France and Spain and began to interbreed with the Spanish white-headed duck; the Spanish complained that it was considered a threat to the latter's genetic integrity, the Royal Society for the Protection of Birds (RSPB) took up the Spanish cause, and in 1999 government commenced (against the advice of English Nature, its official conservation advisors) a trial 'to establish the feasibility of eradication' by shooting ruddy duck in three regions for a three-year trial period. By March 2002, they had spent £900,000 killing 2,558 birds, and ending up with more ruddy ducks than they believed they had when they started. A decision to proceed to a campaign to eradicate the ruddy duck followed in 2005, and by the spring of 2008 the population had been reduced to under 500 birds, though by then nearly as many again were thought to be breeding in France and elsewhere in mainland Europe. The aim is elimination, but the cost, not of elimination but of reduction to under 200 birds in the UK, was thought in 2005 to lie at between £3.6 million and £5.4 million.

Since the shooting takes place in the breeding season and immediately afterwards, the threat to other British biodiversity by disturbance of the breeding lakes is considerable, but brushed aside. There is opposition

---

[22] HMSO, *Biodiversity, the UK Action Plan* (London, 1994), p. 175.

from animal welfare organisations and some local populations, notably at Wigan in Lancashire, but spokesmen of the government Central Science Laboratory describe the opposition as 'much less than expected', and it is also brushed aside.[23] The ruddy duck's offence, like that of the Sika deer, is to make love not war, though the proponents of the cull have been known to describe the male's forceful courtship technique as rape. The journal *British Birds* became a theatre of serious, if impassioned, debate among both scientists and amateurs on the ruddy duck and other alien species, its volume for 2000 alone containing six contributions on the subject.[24]

There is, as already emphasised, general scientific agreement that the spread of non-native species, worldwide, is one of the greatest threats to international biodiversity. But it must be a real question as to where it is appropriate to draw the line, and the blanket condemnation, irrespective of the different degree of threat in different places and to different groups, arouses misgivings. There are several ironies and disturbing facts about the current situation in the UK, of which the greatest is perhaps the prominence of the concept 'genetic integrity'. Is it surprising that minority ethnic groups find the conservationists' view at best insulting and at worst threatening? If the concept of 'genetic integrity' is held to be applicable to every other species, how long will it be before it is held to be applicable to the human species? Conservation organisations sometimes wonder why they have so few members from ethnic minorities, why environmentalism is so often a socially exclusive activity. But the language of the conservation scientist can sometimes sound, to the impartial ear, in danger of coming close to the neo-fascist. 'There is a kind of irrational xenophobia about invading animals and plants that resembles the inherent fear and intolerance of foreign races, cultures and religions', noted one biologist, J. H. Brown.[25] And the language of the journalist who has been listening to the conservationists is even more loaded. To give an American example, a North Carolina newspaper recently associated the decline of the eastern bluebird in the USA in the 1960s and 1970s (actually due mainly to insecticide problems) to the advance of the 'ruffian immigrants', European starlings and house sparrows, and quoted a bluebird lover who called the sparrow the Darth Vader of the avian world.

---

[23] I. Henderson and P. Robertson, 'Control and eradication of the North American Ruddy Duck in Europe', in G. W. Winter, W. C. Pitt and K. A. Fagerstone (eds), *Managing Vertebrate Invasive Species* (USDA National Wildlife Centre, Fort Collins, 2007).

[24] M. Avery, 'Ruddy ducks and other aliens', *British Birds*, 93 (2000), p. 500; C. Bibby, 'More than enough exotics', ibid. pp. 2–3; D. Goodwin, 'Introduced birds', ibid. pp. 501–4; T. Lawson, 'Ducking the real issues', ibid. p. 93; K. G. McCracken, J. Harshman, M. D. Sorenson and K. P. Johnson, 'Are ruddy ducks and white-headed ducks the same species?', ibid. pp. 396–8; B. Zonfrillo, 'Ruddy ducks', ibid. pp. 394–6.

[25] Quoted in Budiansky, *Nature's Keepers*, p. 130.

The paper concerned is not important, nor is this an important comment, but its typicality is significant.[26]

The present position is also a mess. Because non-native species are no longer added to the schedules of protected species in Britain, simply because they are alien, some which could do with protection are denied it: the colony of wall lizards on the Isle of Wight is presumed on ambiguous evidence not to be native, but needs protection in northern Europe.[27] There are several Asian pheasants and at least one Asian duck – the mandarin duck – which are certainly introduced but of small natural populations and possibly globally at risk.[28] On the other side of the coin, there are still interesting anomalies where non-native species are given formal or informal protection and benevolent conservation management. The little owl is still on the protected bird schedule. The vole in Orkney is the European common vole, whose nearest relationships appear to be with common voles in the Balkans. It is thought to have immigrated with people about 5,400 years ago, and is highly regarded by local conservationists.[29] The black rat in Britain, an Asiatic species and the vector of medieval plague, now exists in the UK only on three islands – on the Shiants it was saved by the local Scottish Natural Heritage officer from a dose of warfarin that was to be administered to it in the interests of the native Atlantic puffins; he said there were millions of puffins but only a few score black rats in Britain. Recently, English Nature came to the opposite conclusion on Lundy and set about poisoning the rats in the interests of puffins. The sweet chestnut, probably a Roman introduction, is cherished in Kent and coppiced for common nightingales. Contrariwise, the beech, native in the south but not in the north, was the object of conservation attack on Scottish reserves, despite local protest, because its seedlings threatened to overshadow native flora.

The generally hostile and suspicious view of alien species in the wild taken by conservationists and government in its *conservation* mode, exists alongside tolerance of a financially valuable (and, from the animal right's perspective, an often outrageously cruel) import trade into Britain of all kinds of alien species destined for captivity. Most of this trade is legal, and government in its *commercial* mode imposes few restrictions on the movement of

[26] *Durham Herald Sun*, item on bluebirds, 24 March 2001. For an excellent exploration of the situation in USA, see P. Coates, *American Attitudes to Immigrant and Invasive Species: Strangers in the Land* (University of California, 2007)

[27] A. Quayle and M. Noble, 'The wall lizard in England', *British Wildlife*, 12 (2000), pp. 99–106.

[28] Goodwin, 'Introduced birds'.

[29] Yalden, *The History of British Mammals* (London, 1999), pp. 220–3.

such animals into Britain (unless they are endangered in their native lands) and none on their movement around Britain.

There are quite difficult problems about defining an alien species, of two kinds. The first are historical and conceptual. It can be extremely difficult to know if a species arrived here unassisted within the last 10,000 years, especially where fossil remains are few or ambiguous. Is the little mint pennyroyal truly a native or an old contaminant of hay spread across Europe and Britain in fodder wagons and nosebags? If a ruddy duck ringed on the Chesapeake turns up in the Quadalquivir, will this negate the legitimacy of the eradication campaign, since it will prove that this species (like so many American ducks) can cross the Atlantic unaided?[30] The little ringed plover, which arrived in England in the 1940s, undoubtedly came on its own wings but depends completely on gravel pits to survive. It is regarded as a native species and has all the benefits of protection: the wall lizard, which probably used its natural ingenuity to escape from a cage and is equally dependent on an artificial environment, is regarded as alien and has no benefits of protection. It seems a fine distinction. The brown hare has its own biodiversity action plan but is now thought possibly to have been an ancient introduction.[31] It is doing rather well in places. The pool frog was thought to be an introduction but is now considered probably native. Unfortunately, it became extinct in Britain before this revisionism could help it.[32]

The other problems are philosophical and even more basic. The definition of an alien species rests on a view that humanity is not part of nature (a pre-Darwinian concept) and that what is natural and uninfluenced by humans is especially valuable (a Romantic concept). But it is more realistic to see humans as part of nature, even though their power to change their environment gives them a unique position of responsibility and demands of them a sense of reverence for life other than their own. And in Britain the search for what is uninfluenced by human impact is fruitless: the 'End of Nature', in McKibben's sense,[33] occurred here many centuries ago, and the living space of some of our most treasured native species, like the robin, the swallow and the skylark, is hardly less artificial than that of the little ringed plover and the wall lizard. It is perhaps ridiculous that a waterweed introduced from Iceland to a loch in Scotland on the feet of pink-footed geese is regarded as a native species, but algae brought from the eastern United States to a British port in the bilge water of a bulk carrier are regarded as an alien. But this is not, I repeat, to deny that some alien or non-native species

---

[30] Goodwin, 'Introduced birds'.
[31] Yalden, *British Mammals*, p. 128.
[32] Quayle and Noble, 'Wall lizard'.
[33] B. McKibben, *The End of Nature* (New York, 1989).

are capable of inflicting very severe damage on native biodiversity. Possibly that algae will become a terrible nuisance, as so many marine introductions have proved: from the North American slipper limpet destroying the oyster beds of England to the North Atlantic sea lamprey devastating the trout fisheries of the Great Lakes.[34]

Perhaps it is time to reconsider our terminology and to define again exactly what we are so concerned about. First of all, we ought most certainly to be concerned about cruelty to all sensate animals, and the trade in any caged wildlife seems morally indefensible.[35] Why is it permitted in a civilised country? Second, it would seem wise to make it illegal to release into the wild any captive animal, except as part of a carefully researched reintroduction programme, since the possibility of it becoming a danger to other species is high enough to invoke the precautionary principle.

Third, it is worth considering whether to abandon the whole language of alien species, to revert to terms like introduced and naturalised species, if we are to avoid giving justified offence. More basically, if we are trying to preserve the diversity of species within the country, we might consider whether there are not two different categories – non-vermin, the existence of which threatens the continued existence of no other species, and vermin, which either by existing at all or by existing in excessive numbers, do threaten the existence of other species. If the term vermin is itself too loaded (the Nazis were uncommonly fond of it) 'pest' could be used instead. To revert to 'vermin' or 'pest' would still mean hard decisions. They would go against the American mink, against the carrion crow, against the hedgehog in the Hebrides but perhaps not invariably against the sycamore. It would allow unashamed protection (where appropriate) of non-native species that require it.

Lastly, it seems to me very doubtful if there are sufficient philosophical or ethical grounds for concerning ourselves with questions of genetic integrity where two species or sub-species cross in the wild, even if one of these is an introduction. It is impossible to distinguish the arguments for preserving every genotype from arguments for racial purity in human beings, and if there is a clash between scientific and humanistic values, I am for the humanistic ones. But even the scientific arguments for preventing interbreeding in the wild are open to challenge; surely, if the ruddy duck hybrid with the white-headed duck is better able to survive the rigours of the Spanish environment than the pure-bred ruddy duck, the cross will have

---

[34] Elton, *Invasions*, pp. 26–7, 118–20.
[35] There have been more restrictions placed on the import of wild birds following the advent of avian flu since this was written.

resulted in an organism well-suited to its ecological niche. Just this seems to have happened when the New Zealand grey duck interbred with the alien mallard – a 'genetic extinction' according to one scientist,[36] but where is the loss from the perspective of the birds themselves? If, on the other hand, the infusion of new genes does not help adaptation, the hybrid will die out and the original form of the white-headed duck will survive. It is hard to see the problem. Aldo Leopold famously exhorted us to 'think like a mountain'.[37] We could at least try to think like a species.

It is not just the layperson who has a concern. Forrest and Fletcher, commenting on the anxiety to maintain a native stock of Scots pine, observed that:

> When pressed for a definition of the term [genetic integrity] there is generally some reluctance on the part of those employing such terminology to come to the point, although emotive issues connected with the archival 'preservation of our priceless heritage' and perhaps a variety of 'ethnic cleansing' are seldom far from the surface.[38]

They drew the conclusion that not only was the term feeble in scientific terms, but that the pursuit of the ideal would be foolish, since the greater the available gene pool the better placed is any population to respond to stress, such as that currently posed by rapid climate change.

In sum, the way the alien species conundrum has developed leaves conservationists with a number of nettles to grasp. If they begin to do so, they might regain the confidence of a public beginning to be put off by some of the language they use and some of the strange actions they take. Fortunately, there are signs that some scientists are themselves seriously starting to contemplate the problem, and seeing it in new perspectives.[39]

---

[36] Simberloff in Elton, *Invasions*, pp. xii–xiii.

[37] A. Leopold, *A Sand County Almanac* (Oxford, 1996: 1st edn 1949), p. 140.

[38] G. I. Forrest and A. F. Fletcher, 'Implication of genetic research for pinewood conservation', in J. R. Aldous (ed.), *Our Pinewood Heritage* (Farnham. 1995), pp. 97–106, quote on p. 99.

[39] See, for example, *Glasgow Naturalist* (2001), supplement: 'Alien species: friends or foes' – sixteen papers from varying perspectives.

Fig. 11 The harvest field at Comiston, outside Edinburgh, in 1945. The tractor is pulling a self-binding reaper (not yet a 'combine'), that ties the crop into sheaves which will be set upright to dry in triangular stooks. The great farming revolution has begun with the displacement of horses, but the heyday of pesticides and fertiliser subsidy lies in the future. Photographer: R. M. Adam. Courtesy of St Andrews University Library: RMS-H8165A.

Modern Agriculture and the Decline of British Biodiversity*

We are generally told that the greatest threat to world biodiversity in the twenty-first century is global warming. There is no reason to doubt that, but this is about the environmental history of the UK in the immediate past, not a look into the future. In the late twentieth century, biodiversity in Britain suffered unprecedented loss in a short time span. It did so at a time when environmental concern had never been higher, indeed when Britain was famous for the supposed power and demonstrably large membership of environmental voluntary organisations. It is a story of their powerlessness.

The main cause of biodiversity loss was agricultural intensification. The form it took would have been impossible without science. However, it was not an inevitable consequence of scientific discovery or of some unspeci-fied march of progress, still less of globalisation, but rather largely of the reverse, of the distortion of the market by governments in thrall to that most ancient of political forces in Britain, the landed interest, which in no way represented the people at large or even the rural community. As a cause of biodiversity loss, manipulation of the market could yet be outstripped by climate change. It has not happened yet. How could such a thing happen?

Let us first of all establish that it did happen. Biodiversity loss is nothing new; it has been going on for millennia, and in historic time it has often been associated with agriculture: as in the loss of the wolf due to the interests of pastoral farmers, from England in the Middle Ages and from Scotland in the seventeenth or eighteenth centuries; or the loss of the great bustard due to agricultural enclosure between the sixteenth and eighteenth centuries. With the onset of modernisation after 1800, biodiversity loss undoubtedly occurred on a large scale through the drainage of wetlands, the pollution of rivers, the smoke pollution of industrial towns, and the preservation of

* This paper was given at the Kobe Institute, Japan for the Oxford Kobe Seminar, and subse-quently published in T. Mizouguchi (ed.), *The Environmental Histories of Europe and Japan* (Nagoya University, 2008), pp. 99–107.

I am grateful to Guy Anderson of the Royal Society for the Protection of Birds for his help, but he is not responsible for the conclusions.

grouse moors and private woods for game of all descriptions, with the con-comitant destruction of raptors. On the other hand, the impact of the first phase of industrialisation was limited by the character of Victorian farming, with its high organic inputs: it was a good time for starlings, skylarks, thrushes and rooks.[1]

This situation lasted well into the twentieth century and, after 1950, there were significant signs of recovery from some of the specific losses of the previous century. Wetland birds partly recovered and returned, due both to conservation efforts and to the creation of new habitat through flooded mineral extraction pits and water-supply reservoirs. Rivers were greatly improved in terms of pollution by the 1980s, and fish unseen for a century returned. The smoke plumes of cities were abated from the 1950s onwards, and certain butterflies and lichens driven off in Victorian times came back, very slowly, to old haunts in and around the towns. A changed attitude and bird protection laws, combined with a fall in the number of gamekeepers in most parts of Britain, led to a partial, yet for some species, dramatic, recov-ery in raptor numbers. Some conservation reintroduction schemes have been dramatically successful, such as those of the red kite and large blue butterfly. All this was good news.

There were also some later twentieth-century developments that were expected to have very adverse effects on biodiversity, such as the increase in leisure and the spread of towns. Some did. The unhindered use of beaches, especially by unleashed dogs, did considerable damage to the breeding bird populations that used them, such as little terns and ringed plover. But the increased use of the wider countryside through day visits and improved access had none of the catastrophic impact some conservationists had feared in the 1940s,[2] despite an enormous increase in car ownership from 200 per thousand of the population in the 1970s to 400 per thousand today. And the spread of towns, although it led locally to very considerable loss of heathland habitat and increased disturbance on the heaths that remained for rarer species in the south of England, often produced gardens, which, however unremarkable in themselves, turned out to be richer in common wildlife than the surrounding farmland.[3] The collapse of rustbelt industry left oases of brownfield land unpolluted by insecticide drift, where invertebrates (even

---

[1] M. Shrubb, *Birds, Scythes and Combines: a History of Birds and Agricultural Change* (Cambridge, 2003); J. Winter, *Secure from Rash Assault: Sustaining the Victorian Environment* (Berkeley, 1999); R. Lovegrove, *Silent Fields: the Long Decline of a Nation's Wildlife* (Oxford, 2007).

[2] For example, Seton Gordon, *The Highlands of Scotland* (London, 1951).

[3] K. J.Gaston *et al.*, 'Gardens and wildlife, the BUGS project', *British Wildlife*, 16 (2004), pp. 1–9.

quite rare ones) often flourished. In these respects, increases in national prosperity, or the 'march of progress' as the more optimistic age of the 1960s termed it, did not (and does not) seem entirely incompatible with a rich biodiversity.

However, when we turn to farmland an entirely different picture presents itself. Farmland in Britain comprises 76 per cent of the land surface, a higher percentage than in any of the other twelve members of the European Union (as it was comprised at the end of the twentieth century) except Ireland; so what happened on farmland was absolutely critical to what happens to biodiversity in the UK as a whole.[4]

Let us consider birds first, as they have been the most closely studied group. The first crisis occurred in the late 1950s with the arrival of the new organochlorine pesticides, DDT as a crop spray and dieldrin, aldrin and heptachlor as seed dressings. Particularly in the cereal areas of the east of England, but for raptors more widely, there was a dramatic decline in about twelve species of birds associated with farms. However, it did not last. Following first a voluntary ban on these chemicals in 1961, and legally binding measures subsequently, numbers of most species (though not grey partridges or rooks) bounced back. The British Trust for Ornithology's Common Bird Census, which also began in 1961, showed broad recovery or stability during its first decade. Then, between 1970 and 2004, populations of farmland birds fell by 40 per cent; most of that decline was concentrated between 1978 and 1993.[5]

A broadly similar picture shows in other groups. Insects are fundamental building blocks of the wider ecosystem. Out of twenty-five native species of bumblebees, since 1960, three have become extinct and a further nine suffered serious range decline.[6] Out of fifty-nine resident species of butterfly, since the nineteenth century, five have become extinct and twenty-nine others have disappeared from part of their range – much of the loss occurred after 1970.[7] The larger moths declined by a third in numbers between 1968 and 2002.[8] Botanists are no less concerned by the declines of vascular plants, especially those of specialised habitat, and eight out of ten of the species showing most decline are ancient weeds of arable fields.[9]

[4] European Commission Directorate-General for Information, *Our Farming Future* (Luxembourg, 1993), p. 36.
[5] Shrubb, *Birds, Scythes*, pp. 181–9; Royal Society for the Protection of Birds, *The State of the UK's Birds, 2005*.
[6] T. Benton, *Bumblebees* (London, 2005).
[7] J. Asher *et al.*, *The Millennium Atlas of Butterflies in Britain and Ireland* (Oxford, 2001).
[8] R. Fox *et al.*, *The State of Britain's Larger Moths* (Butterfly Conservation, Wareham, 2006).
[9] C. D. Preston, D. A. Pearman and T. D. Dines (eds), *New Atlas of the British and Irish Flora* (Oxford, 2002), p. 39.

Not all the decline in biodiversity in late twentieth-century Britain was due to agriculture, but much was indeed directly attributable to new farming practices and to habitat loss and fragmentation engendered by agricultural change. Between 1947 and 1980, Britain lost 95 per cent of its lowland herb-rich grassland; 80 per cent of chalk and limestone grassland; 45 per cent of limestone pavements; 50 per cent of semi-natural woodlands; 50 per cent of lowland marshes and fens; over 60 per cent of lowland raised bogs and heaths; and a third of all upland grassland, heaths and mires. One third of all lowland rivers were altered by drainage schemes, and about a fifth of hedges were destroyed.[10]

Meanwhile, what was the general population doing? Among other things, it was joining conservation bodies in ever-increasing numbers. The National Trust had 20,000 members in 1950, 250,000 by 1971 and 3,400,000 by 2000. The Royal Society for the Protection of Birds had 8,000 members in 1950, 67,000 by 1970 and 1,000,000 by 2000. Although the National Trust's concern was with the preservation of buildings and landscapes rather than of biodiversity *per se*, the RSPB's remit was precisely for the conservation of birds and their habitat. By the end of the twentieth century, it had become the largest and most powerful charity in Europe devoted to the protection of the natural world. And it had certainly had successes, notably in its programmes for the recovery of wetland birds and raptors. Despite two important books in the 1980s, which could have acted as catalysts, it was remarkably slow off the ground in confronting the challenge posed by agricultural change: not until the 1990s did this reach the forefront of its concerns.[11]

Perhaps part of the reason for this was the circumstances and resolution of the organochlorine pesticide crisis of the 1960s, already touched upon. Here it had been the government-funded Nature Conservancy and not the voluntary bodies that had taken the lead that eventually led to the legal banning of DDT, aldrin and dieldrin in a series of steps between 1963 and 1973.[12] At this time, the main concern for the RSPB had focused on the damage done to raptors at the head of the food chain and, as a legal ban on organochlorine seed dressings had ultimately been followed by substantial recovery both in raptor and in farmland birds, the RSPB and others were largely lulled into thinking that the problem had subsided. In fact, as the work of

---

[10] J. Martin, *The Development of Modern Agriculture: British Farming since 1931* (Basingstoke, 2000), pp. 173–4.

[11] K. Mellanby, *Farming and Wildlife* (London, 1981); R. J. O'Connor and M. Shrubb, *Farming and Birds* (Cambridge, 1986).

[12] N. Moore, *The Bird of Time: the Science and Politics of Nature Conservation* (Cambridge, 1987); J. Sheail, *Pesticides and Nature Conservation: the British Experience, 1950–1975* (Oxford, 1985).

Volume of output

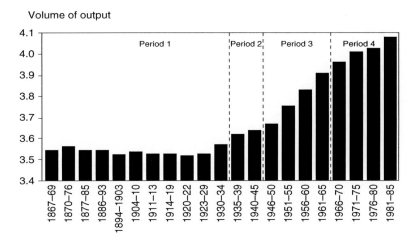

Chart 1 Changes in the volume of agricultural output (measured in billions of pounds at 1986 values) in Britain, 1867–1985. From Paul Brassley, 'Output and technical change in twentieth-century British Agriculture', *The Agricultural History Review*, 48 (2000), pp. 60–84. Reproduced with permission.

the British Trust for Ornithology in particular demonstrated, the incidence of mass deaths and breeding failure among birds had indeed been curbed, but a whole range of other changes to habitat and to farming systems affecting habitat accelerated, and the damage was not at all contained.

But this is to run ahead. What exactly did happen to British agriculture between the end of the Second World War and the start of the new millennium? Chart 1 shows the transformation in volume of output in graphic form. Beginning with the start of government assistance in the 1930s, continuing through the Second World War and accelerating rapidly after 1945, the graph does not go beyond the early 1980s, but it is enough to illustrate the scale of change. Tables 1 and 2 show how, between the late 1930s and 1997, wheat production grew ninefold, barley production tenfold, though oats dropped by two-thirds as horsepower was replaced on farms by tractor power. Barley became more important than wheat between 1955 and 1980, but the 1980s were a decade when wheat easily drew ahead. Using a slightly different baseline comparing 1946 to 1985 and then 1995, milk production almost doubled, poultry production rose twenty-fivefold, the production of beef doubled until the BSE crisis, pigmeat rose sixfold and sheepmeat nearly threefold.

And how was this immense increase in output achieved? It was through a mixture of mechanisation, chemicalisation, breed improvements and management change. In 1936 there were a million horses working on British

Table 1 Cereals grown in the UK (in thousands of tonnes)

Year	Wheat	Barley	Oats
1935	1,771	774	1,849
1950–4	2,576	2,184	2,711
1960–4	3,308	5,891	1,705
1970–4	4,994	8,707	1,174
1980–4	10,651	10,511	556
1990–4	13,845	7,235	509
1997	15,020	7,820	575

Source: J. Martin, *The Development of Modern Agriculture* (Basingstoke, 2000), p. 97

Table 2 Livestock produce in the UK

Year	Milk million litres	Poultry 000 tonnes	Beef 000 tonnes	Pigmeat 000 tonnes	Sheepmeat 000 tonnes
1946–7	8,137	56	584	170	153
1985	15,560	874	1,147	989	315
1995	14,015	1,402	998	1,017	403

Source: J. Martin, *The Development of Modern Agriculture* (Basingstoke, 2000), p. 97

farms, by 1945 there were 436,000, by 1965 fewer than 20,000, and now there are none. Instead, by 1986, there were half a million tractors, and more than half of these were of over fifty horsepower each. The harvest, which in the 1930s had been brought in by a horse pulling a reaper and binder, was now garnered in a fraction of the time previously taken, by a combine harvester: in 1942, there were fewer than 1,000 combines, by 1965 there were 58,000, and they got larger. As they did so, fields were enlarged by removing trees and hedgerows to accommodate them – between 1946 and 1954, annual loss of hedgerows amounted to about 1,300 kilometres a year, accelerating between 1984 and 1990 to 9,500 kilometres, the latter being equivalent to 1.1 per cent of hedgerows destroyed each year. It also became drier: almost 2 million hectares of land were tile-drained between 1940 and 1980, much of it in the 1970s when drainage was often concentrated on wet grasslands of high conservation value.[13] People, with reason, began to speak of arable eastern England becoming a prairie.

In terms of chemical inputs, the problem of highly toxic pesticides in the 1960s has already been discussed, but government action against the worst

---

[13] Martin, *Modern Agriculture*, pp. 16, 106, 174; K. Blaxter and N. Robertson, *From Dearth to Plenty: the Modern Revolution in Food Production* (Cambridge, 1995), pp. 27, 57; Shrubb, *Birds, Scythes*, pp. 150–5.

of these did nothing to reduce the overall uptake of agrochemicals. In 1944, there had been sixty-five approved pesticide products, by 1980 there were over 800, derived from more than 200 synthetic ingredients; expenditure on pesticides peaked in the mid 1980s, but the area treated continued to rise.[14] The main effect of this second generation of chemicals was completely to alter cropping regimes. In place of the three-year ley, which had dominated arable husbandry between 1945 and about 1970, now, by using pre-emergent herbicides with abundant fertiliser, it was possible to keep land under constant cultivation without a fallow break, to abolish rotations, and grow the same crop, such as wheat, year after year. Weeds (and so bird food) became invisible and annihilated under this regime.[15]

Fertilisers there were in abundance. The use of nitrogen fertiliser increased 20-fold between 1930 and 1980, phosphorus 2.5-fold and potassium 6-fold.[16] Inputs since that time have stabilised or slightly dropped, but a consequence has been a serious problem of diffuse pollution of rivers and aquifers by nitrates over large areas of lowland Britain, and the destruction of water quality through eutrophication in many important wetlands. Nitrate concentrations in ground water from fertilisers and animal wastes, in 1997, exceeded the EU guide level of twenty-five milligrams per litre under more than 85 per cent of agricultural land in the European community.[17] Plant breeding substantially improved the yields of cereals, potatoes and grasses, often by choosing strains capable of assimilating and withstanding the effects of the new chemicals, which was one of the main sources of productivity gain after 1980.[18]

Meanwhile, on the side of animal husbandry, similar changes were achieved by the advent of factory farming and silage production. Genetic improvements in the quality of dairy cattle, combined with changes in foddering, led to a doubling of milk yields per cow between 1946 and 1985. The fact that the search for improved protein content of the diet eventually went as far as feeding meat and bonemeal to cattle – feeding cows to cows – also eventually led to the disaster of BSE in 1986. Similarly, the veterinary practice of providing antibiotics for diseases that affected milk yield led to increasing immunity to antibiotics in the human community. Meanwhile, an increasing proportion of beef cattle, pigs and poultry were kept indoors. Free-range poultry had accounted for 80 per cent of production in 1951: by

---

[14] Martin, *Modern Agriculture*, p. 102.
[15] Shrubb, *Birds, Scythes*, pp. 183–99.
[16] Blaxter and Robertson, *Dearth to Plenty*, p. 80; Martin, *Modern Agriculture*, p. 104.
[17] European Commission Directorate-General for Information, *The European Union and the Environment* (Luxembourg, 1997), p. 19.
[18] Martin, *Modern Agriculture*, p. 98.

1980, it accounted only for 1 per cent, though it increased again to 12 per cent by 1990 due to concerns about salmonella and animal welfare.[19] Sheep alone continued to enjoy a relatively traditional outdoor existence on the rough pastures and moorlands of Britain, but higher stocking rates were achieved by additional fertiliser use or simply by overgrazing.

Agriculture: the Triumph and the Shame was the title of an illuminating book by the Conservative MP and farmer, Sir Richard Body. This was published in 1982 before the BSE and foot-and-mouth disasters had hit Britain, and before the income of most farmers, which earlier had risen dramatically, tumbled in real terms back to the depressed levels of the 1930s.[20] The numbers of farmers at the end of the Second World War was about 500,000: twenty years later, this had halved.[21] By the new millennium, 10 per cent of farmers, including the factory-farming 'agribusinesses', accounted for two-thirds of production.[22] The number of farm workers fell by two-thirds or more between 1936 and 1986,[23] and by the end of the century, farming did not account for most of local income even in rural areas. 'The triumph and the shame' is a good summary: the triumph was the unprecedented and extraordinary improvements in agricultural output and the growth of income by a minority of big farmers, the shame the high environmental, social and financial costs of the transformation, and ultimately the failure even to maintain the income of most farmers at a decent level.

Importantly, all these events took place against a background of government intervention. As early as the 1930s, the state had stepped in to help a depressed agricultural sector in a new atmosphere of trade protection and threat of war: marketing boards were established for milk, pigs, potatoes and hops, and subsidies were paid for wheat and sugar-beet production. Farming became exempt in 1929 from paying rates on agricultural land, which in the long run meant that those who polluted the rivers and streams with nitrate or pesticide runoff paid nothing for cleaning them up.[24]

In the Second World War, farmers were widely credited with saving Britain from starvation, as foreign supplies were hampered by enemy submarines: food imports declined by 50 per cent and net farm incomes

---

[19] Ibid. pp. 110–26.

[20] J. K. Bowers and P. Cheshire, *Agriculture, the Countryside and Land Use; an Economic Critique* (London, 1983), pp. 82–3; Martin, *Modern Agriculture*, pp. 150, 201; Blaxter and Robertson, *Dearth to Plenty*, p. 33.

[21] Martin, *Modern Agriculture*, p. 130.

[22] H. Newby, *The Countryside in Question* (London, 1988), p. 10.

[23] Blaxter and Robertson, *Dearth to Plenty*, p. 27; R. Body, *Farming in the Clouds* (Hounslow, 1984), p. 43.

[24] Bowers and Cheshire, *Agriculture*, pp. 97, 115.

tripled. In fact, the part played by the farmers in increasing production was probably less than the part played by the state in regulating output, distribution and pest control, but it was the farmers who got the credit. It was an important time because officials began to accept that the future lay in large-scale industrial farming, government subsidy and a cosy relationship between the National Farmers Union and the Ministry of Agriculture. In the words of a recent commentator, 'the decisions of policymakers during the war helped get British farming to where it is today: a demoralised industry, dependent on subsidy, and responsible for contributing less than 0.5 per cent of GDP'.[25]

After the war, the Labour government passed the 1947 Agricultural Act, which perpetuated the principles of state help for farming and formed the basis of a subsidised regime for the next quarter of a century. They did so because they were faced with ongoing food shortages, a severe balance of payments crisis, and the expectation that rural voters, traditionally Conservative or Liberal, would turn to the Labour party in gratitude. The legislation committed the state to purchasing most agricultural products at fixed prices, determined after consultation with the National Farmers Union, which consequently rose to a position of unparalleled influence. In 1953, this system was changed by a Conservative government to one of relatively free market trading, but only after guaranteed minimum prices for the farmers had been fixed, the deficiency payments being met from Exchequer funding. The NFU continued its privileged position as statutory consultee.[26]

In 1973, Britain joined the Common Market that was, before the end of the century, to metamorphose into the European Union: in doing so, it handed final control over agriculture to the European Commission, carrying out the Common Agricultural Policy (CAP). The Byzantine complexities of the CAP were intended, like the purely British system before, to support agriculture by intervention in the market. Previously the method of subsidy had been to support farmers from general taxation, via deficiency payments: now it became a matter of individual commodity price support regimes and import levies, plus direct production subsidies for hill cattle and sheep. But that did not mean that the British government was powerless: the annual round of negotiations at Brussels took place with the minister closely briefed, as before, by an alliance of civil servants and the farmers' representatives, and the overall aims of the NFU to support intensive production and the

[25] B. Short, C. Watkins and J. Martin (eds), *The Front Line of Freedom: British Farming in the Second World War* (British Agricultural History Society, 2006). Quotation from a review by E. Gill of the above in *Economic History Review*, 60 (2007), p. 844.

[26] See Martin, *Modern Agriculture* and Bowers and Cheshire, *Agriculture*, for accessible accounts.

larger, 'most efficient' farmers were achieved. By the 1980s this was leading to gross overproduction, to export subsidies that led to damage to farming abroad and offended the World Trade Organisation, to high consumer food prices in Europe and to inordinate expense. By 1988 it was taking 63 per cent of the Community's budget to run the CAP; though this had fallen to 58 per cent by 1992, the actual expense grew in those five years from 26 billion to 36 billion ecu.[27]

Protests, led by the USA in the 1985 Uruguay round of talks on world tariff reform, had an effect: slowly a process of partial reform was set in train. In 1992, a package of reforms was agreed that would only become fully implemented by 1997.[28] It was seen that subsidies on the scale of the past could no longer be afforded, either financially or politically. The guaranteed price for cereals and beef was heavily reduced, and the butter mountains and wine lakes disappeared. Farmers were paid to take an annual proportion of their arable land out of production as 'set aside', mainly to help further the reduction of surpluses, but additionally to help offset some of the environmental damage that was now widely seen to have been caused by the old system. Nor were the reforms without consequence: there is evidence from the British Trust for Ornithology censuses of farmland birds that the advent of set-aside helped to stem further decline in several species in the 1990s, though it hardly reversed what had already happened.

In the new millennium, CAP reform was to be carried much further by decoupling subsidies completely from production and paying farmers to maintain their land in good environmental condition, but at the same time set-aside was abolished – a fundamental and complex revolution with uncertain outcomes, but one that carries the subject beyond our self-imposed time limit.

The twentieth century ended in Britain in economic depression for farmers: the reforms' effect was coupled with the catastrophe of BSE and the subsequent collapse of beef sales abroad, and a strengthening pound that further undermined the competitiveness of food exports. Farm incomes 'were sent into freefall': in 1997 they had declined by 47 per cent in real terms.[29] Throughout this period, the British government's ministries of agriculture (of varying names) had negotiated at Brussels with the NFU at their elbow. Small wonder that by the end the NFU was losing members

[27] European Commission, *Our Farming Future*, pp. 13–14.

[28] M. Pollard, *The Engines of European Integration* (2003), pp. 268–71; European Commission Directorate-General for Information, *How Does the European Union Manage Agriculture and Fisheries* (Luxembourg, 1996), pp. 8–11; European Commission, *Our Farming Future*, pp. 21–4.

[29] Martin, *Modern Agriculture*, p. 201.

to rival start-up farmers' unions who felt their true interests had not been represented.[30]

Would the environmental outcome have been any different if there had been less state support? De-rating agricultural land in 1929, already mentioned, decoupled the farming community from direct financial responsibility for run-off pollution into lakes, streams and rivers, and there were many more particularly damaging forms of help to farmers to come. Agriculture became and remains exempt from fuel-oil duty, 35 per cent of cost to other consumers, encouraging the use of energy-intensive techniques and more complete mechanisation.[31] An industry that now contributes less than half of 1 per cent to GDP contributes up to 7 per cent of national greenhouse gas emissions. From 1951, through production grants, fertiliser usage was subsidised, encouraging overuse, and grants were given for every acre of permanent pasture ploughed and re-seeded with ryegrass. Later in the 1950s, capital grants became available for the purchase of machinery, and for long-term improvement such as grubbing up hedgerows and draining fields.[32] Entry into Europe in 1973 led to the abolition or curtailment of subsidies for fertilisers, hedgerow destruction and some forms of upland ploughing, but field drainage, fencing of moorland and so-called 'grassland regeneration' continued at a higher rate of subsidy in upland areas. Headage payments for sheep in the 'less favoured areas' of the north and west of Britain, intended to alleviate the problems of upland farmers, led to gross overgrazing of pastures.[33] On balance, the CAP made matters worse by extreme encouragement to overproduction. Finally, most research in agriculture and advisory services was publicly funded, and driven by the government, in consultation with the NFU, in the direction of encouraging farmers towards intensification.

A particularly distinctive thing about British agricultural policy between the Second World War and 1997 was that it was driven by a desire to replace relatively extensive, low-input husbandry by intensive, high-input husbandry, and the subsidy system did not reward farmers unless they went down this road.[34] This created much higher outputs per acre and per unit of labour employed, but had no social objectives (as in France and Germany) of maintaining the small farmer on the land. Those able to take most advantage of it were the large farmers of the arable counties

[30] Séan Richard, 'The farming lobby: a waning power?', in Berkeley Hill (ed.), *The New Rural Economy* (Institute of Economic Affairs, 2005), pp. 206–18.
[31] Bowers and Cheshire, *Agriculture*, p. 97.
[32] Martin, *Modern Agriculture*, pp. 81, 105; Blaxter and Robertson, *Dearth to Plenty*, p. 80.
[33] Bowers and Cheshire, *Agriculture*, pp. 96–7.
[34] Ibid. pp. 115–19.

of Britain, and the small farmers everywhere went to the wall. When the CAP came along, the big farmers with high production records were also particularly well able to take advantage of a system geared to rewarding production irrespective of demand: the 'barley barons' and their kin did exceptionally well, the small farmers and the natural environment exceptionally badly.[35]

Birdlife decline associated with farming practices pursued under the Common Agricultural Policy has been common across Europe. Agricultural intensification, involving loss of hay meadows, use of inorganic fertilisers, crop monocultures or larger field size, was implicated in 42 per cent of instances of species decline in Europe between 1970 and 1990, and the effect of pesticide use was implicated in 23 per cent of such declines. But it did not have to be that bad. That there was something especially unfavourable to wildlife about the British system is indicated by a contrast with Denmark, a fellow member of the European Community from 1973. Certainly, there were heavy losses to the natural heritage in Denmark as well in the second half of the twentieth century, and these too were often attributable to state-supported agriculture. But a study of farmland birds shows that in Denmark between 1983 and 2001, of twenty-seven bird species associated with farmland habitat, five declined, ten were stable and twelve increased, whereas in Britain over the same period fifteen declined, eight were stable and four increased. Although it was not possible to establish causality beyond doubt, in Denmark over these years pesticide and inorganic fertiliser use declined (partly due to Danish taxation policies), in contrast to Britain, and organic farming increased more than in Britain. It is an interesting indication that farming may be set back on to a wildlife friendly track in an advanced economy, given the will.[36]

How did the British farmers get away with it? First, there was a close relationship between the big farmers who were leaders of the NFU, the civil servants of the agricultural ministries and the government ministers themselves, with the Country Landowners Association playing a back-up role. It was all reinforced by the political power of the House of Lords (where the hereditary peers dominated), and by the commercial muscle of the agrichemical industry. They were quite open about it. Here is an extract from the *Farmers Weekly* in 1984, which, after discussing more conventional forms of lobbying, went on:

[35] Newby, *Countryside*, pp. 9–15; Martin, *Modern Agriculture*, pp. 147–51.
[36] G. M.Tucker and M. F. Heath, *Birds in Europe: their Conservation Status* (Cambridge, 1994), p. 46; A. D. Fox, 'Has Danish agriculture maintained farmland bird populations?', *Journal of Applied Ecology*, 41 (2004), 427–39.

There is the more undercover style of lobbying which the NFU does so well. This means knowing who the most powerful people are on a certain issue and bringing them round to the farmers' point of view. It is the kind of lobbying that goes on at lunchtimes, in bars and at private dinner parties where Cabinet Ministers are mellowed by good claret and port. It's expensive and exclusive, but it works.[37]

Sir Richard Body, himself a Tory member of the Reform Club, described how he used to encounter the Director General of the NFU and the Permanent Secretary to the Ministry of Agriculture, alongside several others 'of considerable influence in agricultural policy', lunching together in 'the very room where the Cobden Club was founded to resist a return to the Corn Laws.' And he remarked on the regard with which one of the agrichemical industries 'government affairs officers' was regarded by the farming establishment: 'As he went about the room, heads would turn and in each knot he joined he became its centre. A stranger to the scene would have assumed he was the most important man there in British agriculture.'[38]

It is evident that some ministers of agriculture felt themselves to be patronised by the farmers' representatives, but Labour and Conservative alike put up with it because of the impression that the farming vote could deliver constituencies at general elections. Estimates varied, but in the late 1940s there were about 100 constituencies where the farming vote was thought to be significant, and as late as 1966 there were sixty constituencies where the number of farmers and their wives was greater than the majority enjoyed by the MP. There were also, incidentally, ninety MPs in 1970 who themselves were farmers, seventy-nine of those Conservatives.[39]

Gradually, however, the political power of the farming interest began to evaporate. Partly this was because the numbers of farmers and farm workers steadily fell. Partly it was due to an increasing anger by the wider voting public at the erosion of the environment, the beauty and the interest of the countryside due to the behaviour of farmers, and resentment that it was funded from the tax-payers' pockets. The BSE disaster in 1986 was the last straw for many, and the tone of government ministers' remarks now began to change. David Curry, junior agricultural minister in the Conservative government in 1991, remarked to Parliament in a way unthinkable a decade before: 'It is not the policy of this government or any hypothetical

---

[37] Quoted in Body, *Farming in the Clouds*, p. 100.
[38] Ibid. pp. 21–2.
[39] G. K. Wilson, *Special Interests and Policymaking: Agricultural Policies and Politics in Britain and the United States of America, 1956–70* (London, 1977), pp. 19–28.

government to allow farmers to become the recipients of perennial social security payments merely because they are farmers.'[40]

But it was too late. As John Gummer, a former Conservative agricultural minister, also reminded the House of Commons in this debate, 80 per cent of agricultural spending in Britain now originated in Brussels, and Britain could do nothing without her partners.[41] The century did indeed end with CAP reform on the table, though not thoroughly achieved. It also ended with much of Britain's biodiversity in ruins. However you read it, it is not an edifying story.

One last point demands attention. If Britain had not gone down the road of domestic subsidy and European protection and manipulation, would there have been biodiversity loss? Of course there would. Almost all that made the destruction possible through chemical, genetic and mechanical innovation would still have been available, since most of it was American anyway. All of it would have been wanted and used by farmers to the extent that they could afford it, and been able to make money by using it. Mostly it would have been the big farmers who could have afforded it, and this fact alone would have provided an impetus for larger farmers, larger fields and more industrial farming. Mostly it would have been farms on good land well placed for the market that would have used it, which implies especially the traditional, largely arable areas most in touch with the dense urbanisation of the south east. Here intensive farming would still have taken place, includ-ing chemical-drenched grain husbandry, but also niche fruit and vegetable farming, and factory farming for poultry, beef, pork and milk. But volumes (and so proportionate damage) would have been smaller as a real market would have to have been found for all the produce. The wise farmer would have had a more mixed farm (so with more habitat diversity) to protect himself against more uncertain markets.

There would have been other significant differences, too. Given the much higher costs of inputs, the successful farmer would have had to target them much more carefully, and Danish experience shows that this is likely to have given rise to much less biodiversity loss. Overgrazing of the uplands to produce unwanted animals would not have occurred, and much of the damaging destruction of traditional meadows and pasture would have been avoided. Possibly small farmers would still have gone out of business, though a greater incentive for organic specialisations might have saved some of them. Land in the north and west would often

[40] P. Baines, 'Parliament and the Common Agricultural Policy, 1990–2', in P. Giddings and G. Drewry (eds), *Westminster and Europe* (London, 1996), p. 203.
[41] Ibid. p. 202.

have gone out of farming and reverted to bog or scrub, as it is doing now. This would have resulted in some biodiversity loss, and a different biodiversity gain. But it would not have been the unmitigated disaster that in fact occurred.

Fig. 12 The wild white cattle in Cadzow park, Lanarkshire, in 1898. This ancient breed was described in the sixteenth century as having 'crisp and curland mane, like feirs lionis', but in the nineteenth century the last herd was kept to graze the medieval wood pasture of the Dukes of Hamilton. Now extinct, their nearest relatives are at Chillingham in Northumberland. Photograph: Valentine Collection. Courtesy of St Andrews University Library: JV-29737.

CHAPTER 12

History, Nature and Culture in British Nature Conservation*

REGARDENERS AND REWILDERS

It is a well-known and well-grounded generalisation that there are funda-mental differences in attitudes to nature between America and Europe. In America, the common view is that true nature is undisturbed wilderness, and conservation at its most ambitious is about 'rewilding', about return-ing to the primeval wonderland before man spoiled it. It has been so since the days of Emerson and Muir.[1] In Europe, by contrast, nature is not seen as apart from human impact. Since Classical times the most admired land-scapes have been those that bear the imprint of man's hand in patterned fields, tended woods and managed moorland. Conservationists in Europe, when they try to restore biodiversity losses, try what might be described as 'regardening', returning land to some previous state where human interven-tion seemed to have been more ecologically benign.[2]

Like all generalisations, however, it is both broadly true and grossly over-simplified. William Cronon is one American environmental historian who is critical of those who imagine that nature was ever divorced from man, or that the recovery of untouched wilderness can ever be fruitfully pursued as a goal.[3] Many practitioners of ecological restoration in the USA find them-selves repairing damage to biodiversity while also respecting old cultural landscapes and their archaeology. Similarly, in Europe, there are both scien-tific and romantic admirers of 'wild land' who trace their intellectual roots back to Wordsworth, Walter Scott and early nineteenth-century pursuits of

* This is a considerably revised version of a paper given to a conference on attitudes to land res-toration, 'Restoria', at the University of Zurich in August 2006, organised by Marcus Hall.
[1] R. F. Nash, *Wilderness and the American Mind* (New Haven, 1967); M. Oelschlaeger, *The Idea of Wilderness* (Yale, 1991).
[2] M. Hall, *Earth Repair: a Transatlantic History of Environmental Restoration* (London, 2005); K. Olwig, *Nature's Ideological Landscape* (London, 1984).
[3] W. Cronon (ed.), *Uncommon Ground: Rethinking the Human Place in Nature* (New York, 1995): the opening chapter, by Cronon himself, is tellingly titled 'The trouble with wilder-ness, or getting back to the wrong nature'.

the 'sublime' and the 'terrific', as well as to influences from the USA. Even the language of rewilding is becoming fashionable in Britain, as we shall see.

The prime example of a European rewilder is surely Franciscus Vera, the Dutch ecologist whose views on woodland history and management have had considerable influence in Britain and elsewhere, as well as in the Netherlands itself.[4] For Vera, nature exists independent of culture and was at its most perfect before humanity interfered by the intervention of farming. For him there was an 'original-natural' state that forms a baseline which we need to understand if we are correctly to direct our attempts to restore natural processes and achieve a rich biodiversity. For him, nature conservation is about process. It is not about fiddling with little nature reserves to which one or two endangered species have retreated, or turning the clock back to one particular century. It is about understanding a time when human beings were without history, certainly without agriculture, in order to recover and release forgotten process. In the context of his own interests, it is about using large herbivores to create a landscape of groves, glades and wood pasture, which he believes was the true baseline for an 'original-natural' Europe.

His name is inescapably linked with Oostvaardersplassen, the great Dutch reserve closely associated with his work and ideas. The Netherlands, the land of tulips and the home of gardening, does not have much wild land, however defined. In 1975, an area of 3,600 hectares of land recently reclaimed from the sea was declared a nature reserve, the biggest in the country. An additional 2,000 hectares was added in 1982. The land was not restored as tidal mudflat or seabed, but became the boldest experiment in terrestrial rewilding in Europe. In the words of the official book on the project:

> Here nature conservation got a new dimension. No longer conserving existing nature in mostly small reserves but allowing freedom to natural processes, and see what happens. From nature conservation to nature development. Since the seventies the Oostvaardersplassen is the flagship of a new direction. Many people from abroad look at it and are astonished about what appeared to be possible on only half an hour driving from the city of Amsterdam in such a densely populated country as the Netherlands.[5]

From reading this quotation, one might get the impression that the Dutch had enclosed the polder and gone away to leave it to nature without further

---

[4] F. W. M. Vera, *Grazing Ecology and Forest History* (Wallingford, 2000); F. Vera and F. Buissink, *Wilderness in Europe* (Tivion Nature, Netherlands, 2007).
[5] Vincent Wigbels, *Oostvaardersplassen: New Nature below Sealevel* (Zwolle, n.d.): quotation from preface by Adrí de Gelder.

intervention. Nothing could be further from the truth. Under the guidance of Vera, Oostvaardersplassen is now home to thousands of introduced red deer, Heck cattle and Konik horses, which roam the reserve untended, unculled, unfed and unhindered. According to Vera's theory, which has proved immensely stimulating to conservation biology throughout northern Europe, this replicates a critical and forgotten natural process of the true original-natural past (i.e. before agriculture), that is to say, the benevolent impact on the land of numerous large herbivores. The Heck cattle are standing in for the lost aurochs of prehistory.

So Oostvaardersplassen is at once regardened and rewilded. It was drained by man to create a terrestrial nature reserve on what was once a seabed and subjected by man to the impact of thousands of introduced wild and domestic animals: regardened. Then it was left utterly to itself and the forces of nature: rewilded. Should it ever be shown that Vera's theory is wrong, and that such herds were not as numerous, large or influential in the pre-agriculture natural state as he argues, the experiment of their introduction would be shown to have been (in its own terms) inauthentic, driven by a process not entirely 'natural'. But that would not for a moment diminish its objective value in creating and maintaining an astonishing biodiversity in this great reserve.

English Nature was attracted by the drama and scale of Vera's vision, and started enquiring whether 'naturalistic grazing' on the model of Oostvaardersplassen could realistically be adopted as a conservation strategy in Britain. They immediately saw that British animal welfare legislation, among other legal and bureaucratic obstacles, would render non-intervention techniques of herd management inoperable: in other words, culture would obstruct nature (if nature it be). There were further reservations. It is not so clear to English Nature as it is to Vera that naturalistic grazing in a lowland context would be the way back to a pre-Neolithic wild landscape, or indeed what kind of landscape it might create. It could have unforeseen and undesirable effects on existing sites of conservation value. So they recommended something much less radical, the experimental and managed use of large herbivores on networks of enlarged and linking reserves existing in a matrix of agricultural countryside, in order to preserve what is there and make a better habitat for what might naturally arrive. It is very much a regardener's option, seen as a pragmatic way forward for an organisation set up by government to conserve natural heritage under the constraints of national and European law.[6]

---

[6] K. Hodder and J. Bullock, 'Nature without nurture', unpublished paper given at the 'Restoria' conference in Zurich, 2006.

It is interesting to put this debate in the context of W. M. Adams's critique of modern British conservation policy. It was George Peterken who coined in 1981 the phrase 'future nature', when he made a distinction between five 'qualities of nature'. He began with 'original nature', the state that existed before significant human impact, Vera's 'original-natural' baseline. He ended with 'future nature', the state that would develop if human influence was completely removed now. It would not be the same as original-natural because soils and climate changed, and are still changing, and because of the introduction of non-native species.[7]

Bill Adams has taken up the phrase in his book, *Future Nature: a Vision for Conservation*,[8] where he criticises British conservation policy as too often faint-hearted, fixated on small preserved sites, ignoring the wider country-side, and reluctant to 'release the wild'. By this he means, like Vera, releasing natural processes to work without constraints, even if this means an element of unpredictability and the risk of losing some familiar forms of nature in the shape of present habitat and biodiversity. The gains would be greater, more natural, than the losses.

But this is vision. Let us take what actually happens in British conservation policy, and see how it uses, or fails to use, history, and then let us see how its confused attitude to history arises from a particular scientific view of the relationship between nature and humankind. Future nature, it will be proposed, has to arrive from acknowledgement of past culture.

History in British Nature Conservation

In Britain, nature conservation has until now been overwhelmingly site based, despite some attempts, notably by Scottish Natural Heritage in the early 1990s to break out of that straitjacket. There were signs at the start of this millennium that this might change with new agri-environment schemes to ameliorate the devastation of the wider countryside, but these are now threatened with infanticide by strangulation from worldwide rises in food prices.

On protected sites, where most of what is valuable is concentrated, there are several current approaches that involve both restoration and an appeal to the past. Large-scale self-proclaimed rewilding is still relatively unusual and mainly consists of projects for the future. In England, the most exciting involves the Wet Fens for the Future Partnership, a consortium of voluntary bodies, private landowners and government organisations that plan to return

---

[7] George F. Peterken, *Natural Woodland: Ecology and Conservation in Northern Temperate Regions* (Cambridge, 1996), pp. 12–13.

[8] W. M. Adams, *Future Nature: a Vision for Conservation* (London, 2003).

over 7,000 hectares of farmland to fenland in East Anglia.[9] In Scotland, the most ambitious current project, which may turn out to be a pipe dream, is to turn 9,510 hectares of Glen Alladale in Sutherland into a fenced reserve with reintroduced elk, wild boar, lynx, bear and wolves.[10] Even that would be dwarfed by Philip Ashmole's vision of returning montane birch and pine to over 6,000 square kilometres of the Scottish mountains above the current timberline, but that is certainly no more than a gleam in the eye.[11]

The term 'rewilding' is undoubtedly in fashion in British conservation, but what it means by those who use it here is another matter. The Wildland Network on its website lists forty-nine current projects described as rewilding.[12] In England and Wales, out of thirty-four sites only four involve more than 1,000 hectares, and several are under 10 hectares: the smallest is 0.13 hectares, barely a garden. In Scotland, thirteen are above 1,000 hectares, including several belonging to the John Muir Trust and other voluntary bodies. Most projects in Scotland are on land that is quite wild already by European standards (Glen Affric, Glen Finglas) and simply aim to improve it for nature conservation. One, the Tweed Rivers Heritage Project, claims to be rewilding over half a million hectares, starting in 1999 and finishing in 2007, but this is surely playing with words. In no way has the pleasant agricultural landscape of the Tweed been suddenly turned into a wilderness: there would be public uproar if it had. Rather, within this large area there has been a certain amount of tree planting and improvement of riverine habitat: admirable, but not the creation of wilderness. Obviously there is less scope for rewilding in Britain than in countries like the USA, where, for example, a project is in train to reclaim for 'buffalo commons' 1.4 million hectares of the grasslands of the Midwest.

Most habitat restoration in Britain, then, has perforce been of a 'regardening' character: less ambitious, small-scale management involving sites of a few dozen hectares or less. It is indeed sometimes derisively called conservation 'gardening' by its detractors, who question whether much that is sustainable can be carried on at this scale. Gardening and not rewilding, however, has been until now the staple approach of the wildlife trusts, of the Royal Society for the Protection of Birds, of most other conservation charities, and of most of the work on the ground by the three government conservation agencies, English Nature (now Natural England), Scottish Natural Heritage and the Countryside Council for Wales. There is, however, a third approach gaining ground, sometimes called 'landscape-scale change' that calls for a

---

[9] Adams, *Future Nature*, pp. 160–74.
[10] See http//www.wildland-network.org.uk/glen_alladale.htm.
[11] Philip Ashmole, 'The lost mountain woodland of Scotland and its restoration', *Scottish Forestry*, 60 (2006), pp. 9–22.
[12] See http//www.wildland-network.org.uk, as consulted on 8 May 2008.

combination of the first and second. It consists of setting up networks of new or restored conservation habitat within a matrix of agricultural land, such as patches of woodland linked by hedgerows or of rough open grassland within insect-flight of each other. It is in some ways closer to rewilding than to gardening and some of the projects in the Wildlands Network database would fall more comfortably under this heading, such as the Border Mires project in the north of England, planned to associate 11,851 hectares of individual mires that vary in size from 2.5 to 400 hectares. The emphasis appears often to be on the abundance rather than the quality of what is created, eschewing micro-management in favour of releasing supposedly natural processes (for example in the work of the Woodland Trust and various agri-environment schemes). However, at least for insects, there is a realisation that exact quality of the patches is critical (exemplified in the various Marsh Fritillary recovery schemes in England and Wales). For the purposes of this chapter, it is convenient to emphasise the contrasts between rewilding and gardening, without forgetting that landscape-scale change can combine the best of both, especially in a long-settled and long-farmed continent like Europe.

Historians and archaeologists are not routinely employed in nature conservation, yet all three approaches to habitat restoration imply a vision of the past, and appeals to the past are often explicit and elaborate. There could be no better example than the controversy around Vera's account of 'original-natural' woodland and the experiment at Oostvaardersplassen, discussed above. This initiative and the theory behind it has had great influence and given rise to great disagreement. Palaeoecologists in Denmark and Ireland, for example, consider Vera's conclusions wrong, and prefer the older paradigm of the Mesolithic forest as predominantly continuous cover, broken by windthrow, small lawns and bogs.[13] English Nature, led by its chief woodland ecologist Keith Kirby, commissioned an impressive report from ecologists, palynologists and specialists in macrofossil molluscs and invertebrates to try to elucidate the 'original' nature of English woods in the context of this controversy. They concluded cautiously with an English compromise:

> Parts of the Atlantic forest may have looked like a modern wood-pasture and there might have been some permanently open areas; but the majority seems likely to have been relatively closed high forest, but with a component of temporary and permanent glades.[14]

---

[13] J.-C. Svenning, 'A review of natural vegetation openness in north-western Europe', *Biological Conservation*, 7 (2002), pp. 290–6; F. J. G. Mitchell, 'How open were European primæval forests? Hypothesis testing using palaeoecological data', *Journal of Ecology*, 93 (2005), pp. 168–77.

[14] K. H. Hodder, J. M. Bullock, P. C. Buckland and K. J. Kirby, 'Large herbivores in the

It is worth noticing two things about how history is used in this debate. First, the debate has been conducted by the scientific community in rigorous historical terms, often by the most sophisticated means available. That it has been carried out using the methods of laboratories does not make it less historical. It is simply that the Mesolithic did not generate documents, but it left a wealth of fossil pollen, snails and insects, and a few bones. Second, there is an unquestioned assumption that the Mesolithic past has the greatest relevance, since that was the last age that can be considered 'original-natural'. What we have around us today, the conservationists often seem to be telling us, is in degraded genealogical descent from a purer, noble, natural, wilder time. The English Nature report, however, is aware that the wildwood might itself already have been substantially changed by the hunter-gatherers, a possibility raised by A. G. Smith as long ago as 1970 and reiterated recently by environmental historians like Ian Simmons.[15] Mesolithic people had the capability, by ringbarking and fire, of modifying woodland to assist both hunting and gathering, by manipulating grassy spaces and increasing scrub and edge in a mosaic within the forest. If they were capable of doing it, they probably would do it, and if they did it on a large scale, it would have implications for applying Vera's model today without additional human management. Much would depend on the size of Mesolithic populations and the stretch of a tribal range, factors likely to vary over time.

Some rewilders, particularly in Scotland, are much less interested than English Nature or than Vera in environmental history, ancient or modern. Once they have convinced themselves of an original-natural state, typically based on a popular myth of the Caledonian Forest, they carry on with recreating what it represents, irrespective of 5,000 years of climate change and human activity, or its mark on the landscape, and certainly irrespective of research. This can lead voluntary bodies and individuals to attempt idiosyncratic landscape change. What they will end up with will not be natural, but cultural landscapes that memorialise early third millennium conservation theories. These will be interesting landscapes in themselves, but probably not what their originators wanted.

Importantly, proponents of rewilding always (and rather paradoxically) deny any intention to recreate a particular point of time. In this they

---

wildwood and modern naturalistic grazing systems', *English Nature Research Reports*, 648 (2005), p. 169.

[15] A. G. Smith, 'The influence of Mesolithic and Neolithic man on British vegetation: a discussion', in D. Walker and R. G. West (eds), *Studies in the Vegetational History of the British Isles* (Cambridge, 1971), pp. 81–96; I. G. Simmons, *An Environmental History of Great Britain from 10,000 Years Ago to the Present* (Edinburgh, 2001), pp. 42–8.

differ from those who seek to restore a historic building, who either want to recreate its appearance at a particular point (say, on construction) or to 'conserve as found', a modern preference for retaining the patina and alterations up to the point of bringing the building into restoration management. Nature conservationists have two emphases in their response to the query about why they do not seek any point in time. Many say that they are trying not to recreate a period anyway, but to liberate a natural process: at Oostvaardersplassen, this process is naturalistic grazing, where ungulates impact on the landscape without human management or intervention. Others would say that you cannot exactly recreate a point in the past, partly because it is unknowable in sufficient detail but mainly because, with the best will in the world, many of the parameters are different now: the absence of wolves or the presence of non-native (alien) species, or simply a different climate where change may or may not have anthropogenic causes. There is the further point that ecosystems, even those minimally disturbed by man, are never static. Clementsian notions of a climax, which, once achieved, was forever stable, have been replaced by a picture informed by chaos theory, of ecosystems in constant change, one tree species randomly replacing another as dominant in the forest, pine or elm advancing or retreating as affected by climate or disease, disturbance the norm. Yet this does not prevent an obsession with the original-natural, taken in Europe to be the Mesolithic, in America the pre-Columbian or at least the time before white colonisation.

However, we can lift the stick from the other end of time. What is the point of harking back to an original-natural that in Europe is unknowable, irrecoverable and 5,000 years distant in time? It is simpler in America, where white colonisation is sometimes barely 150 years old, as on the buffalo commons. The European landscape around us is not Mesolithic and original-natural but modern and cultural. It is impoverished and becoming more so, but the biodiversity we value, and the *Red Data Book* species we especially treasure, are largely the product of the cultural landscapes that immediately predated this one. The terrestrial ecosystems of the European world before 1800, as Paul Warde has demonstrated in respect to the German forests, were determined by societies more anxious to minimise internal social friction and economic risk than to maximise the use of resources, which led to organic economic systems that allowed greater space for biodiversity than ours.[16] Indeed, by favouring a mosaic of land uses to meet as many different needs within one communal territory, it could be argued that they maximised variety of habitat beyond what was there before farming.

---

[16] P. Warde, *Ecology, Economy and State Formation in Early Modern Germany* (Cambridge, 2006).

Historians know, or could find out, a great deal about the cultural land-scapes of the past 400 years or so, what they looked like and how they were maintained, and the file of contemporary information about its biodiversity thickens as we move from the early days of the Royal Society in the seventeenth century, through the Linnean scholars of the eighteenth century, to the immense labours of the Victorian naturalists and their recent successors.[17] It is the conservation gardeners who know this, although in practice they seldom turn to professional historians for help. Because the rewilders work on a grand scale, they are often dismissive of managers looking after a reserve of limited size, considering them to be fiddling at the edges. In many cases, they criticise them for actually working against the dynamics of nature by trying to hold back an ecological succession that would convert a dry heath or a raised bog into a birch wood, or a marsh into a rushy field. In the rewilders' ideal world, the area of ecological restoration would be large enough to accommodate any losses from ecological succession, as there would be space for nature to create new heaths and marshes from the free rein of natural processes.

To this the gardeners retort that we do not in Britain or Europe live in an ideal world, that there is little prospect of doing so, and that without their constant effort on small sites most of the highly valued assemblages of *Red Data Book* flora and fauna would become locally or nationally extinct. They could add, but seldom do, that the vast majority of British sites (and the existence of *Red Data Book* species in them) are only there because the sites are cultural, not natural, and have been maintained by past generations of farmers and others preventing the natural ecological changes that would have overwhelmed them. It is the character of ancient cultural landscapes to create niches, and the nature of biodiversity especially to appreciate them. Peter Marren described the Nature Conservancy in the 1950s as 'fully aware that nature reserves would have to be managed if they were not to dry out, or become overgrown by scrub or rank grass, or harbour the "wrong" sort of wildlife'.[18] It is still true, and the management needs to be based on historical knowledge as well as on science.

One interesting corner of British conservation that almost by definition relates to conservation gardening rather than rewilding is safeguarding the ancient weeds of arable cultivation. Many of these are not native plants but 'archaeophytes', defined by botanists as alien species known to have been present since before 1500, and some of them believed to have been present since Neolithic times. As a group, they have been severely threatened by

---

[17] D. E. Allen, *The Naturalist in Britain: a Social History* (London, 1976).
[18] P. Marren, *England's National Nature Reserves* (London, 1994), p. 23.

changes in late twentieth-century agriculture, especially by herbicides, by intensive farming that obviates a need for a fallow break, and by effective seed-cleaning techniques.

There are 141 (by some accounts 167) species of archaeophytes, of which thirty-nine are among the 100 plants in the British flora to have shown the most marked relative decline since 1930. Eight out of the ten species to have shown most decline are archaeophyte weeds of arable fields.[19] The names of some of the most threatened reveal their long cultural association with people and farming: lamb's succory, corn cleavers, corn buttercup, shepherd's needle, pheasant's eye, corn chamomile, stinking chamomile, corn marigold, black bindweed, corn gromwell, field woundwort, corn salad, Good King Henry, fat hen, Venus's looking glass.

The rarest of them eke out their existence on the edges of fields in marginal places in the south of England, sometimes on the occasional farms that escaped, for one reason or another, the full blast of modern ideas. They are even threatened now in some places by ill-considered agri-environment schemes that favour planting field margins with hedges and trees, or allowing old fields to revert to scrub. And why not let them go, the purist may ask? What has their presence got to do with nature? They came with one farming regime, let them go with another.

That, however, and interestingly so, is not now the view of the UK Joint Nature Conservation Committee (JNCC), whose *Red Data Book of Vascular Plants* (2005) lists a whole swathe of archaeophytes, though no neophytes (i.e. those plants believed to have arrived since 1500). It is easier than it was, but still much more difficult to get protection in Great Britain if you are an alien plant than an alien person, and heaven knows that is difficult enough.

Conscious of breaking new ground in protecting alien species at all on such a scale, the JNCC offered three justifications. First, archaeophytes are, as a group, stable or declining, often throughout the European range, and some of the arable weeds are under threat of extinction in the UK. Second, archaeophytes, unlike neophytes, tend to have a world distribution that is unknown or uncertain, and are often not regarded as native anywhere: to neglect them would be to risk their total loss from the planet. Thirdly,

> archaeophytes are of considerable historical and cultural interest. They have developed (and exploited) a close relationship with man which is, in effect, one of <u>commensalism</u> – many archaeophytes are, quite

[19] C. D. Preston, D. A. Pearman and T. D. Dines (eds), *New Atlas of the British and Irish Flora* (Oxford, 2002), pp. 37–8.

literally, 'followers of man'. The way in which humans now value these species is partly a consequence of having been so intimately associated with them over such a long time period.[20]

So culture as well as nature has here become explicitly a reason for nature conservation. It is to be hoped that a role will now be found for agricultural historians in learning how to tend and recover populations of these plants. And what is true of the plants of arable fields is no less true of the plants of traditional hay meadows and other grasslands, where the expertise of the documentary historian can again usefully be added to that of the ecologist.[21] Thanks to the work of Oliver Rackham, it is already true of the plants of traditional ancient woodland, where the informed maintenance or revival of coppicing can bring the woodland flora back to life, if levels of deer grazing permit.[22] These are all ways of gardening and they all have their critics on that account. But if the need for conservation gardening is accepted, so should be a role for the historian to recount the archaic practices that once maintained the rich cultural mosaic of ecological niches, just as the archaeologist has a role to pronounce on the history of the weeds themselves.

Peter Marren made a relevant point about this when he described the early days of Britain's National Nature Reserves in the 1950s:

There was a great deal to be learned from traditional land husbandry and the craft industry, from coppicing and shelterwood systems in woodland to the commoning practices of rough grazing, turf digging and reed harvesting ... [but] there is little evidence that the scientists talked to the woodmen, the commoners or the thatchers. Theirs was a hermetic world of seminar rooms and laboratories, and they thought in terms of the future rather than the past. As a result, they were in danger of having to reinvent the billhook.[23]

There is still a tendency to rely too much on science and too little on knowledge of past practice, but recent work at Sussex University is exactly designed to bring such knowledge to bear. It is an oral history project exploring how

[20] C. M. Cheffings and L. Farrell (eds), *The Vascular Plant Red Data List for Great Britain, Species Status 7* (Joint Nature Conservation Committee, Peterborough, 2005).

[21] See, for example, the work that John Rodwell is currently engaged on in the Dearn Valley, Yorkshire.

[22] Oliver Rackham, *Ancient Woodland: its History, Vegetation and Uses in England* (Colvend, 2003).

[23] Marren, *Nature Reserves*, p. 24.

older agricultural methods and information might be utilised to manage the streamsides of the upper valley of the Sussex Ouse, for a combination of flood prevention and improved ecological condition, including the restoration of wild-flower meadows.[24]

Many habitats in Great Britain have been entirely formed by cultural manipulation of nature, especially farming and hunting. Many more have come about through crude industrial activity, and the biggest threat to several species of rare insects in Britain is the building over of urban brownfield sites.[25] Among significant habitats, including small nature reserves, are a surprisingly large number involving disused historic mineral extraction sites, some of considerable interest to industrial archaeology. A quick trawl through the files of the journal *British Wildlife* in the last few years identified coal-mining relics in Yorkshire; such reserves as the clay pits near Peterborough, with their nationally important population of Great Crested Newts; metal mines in Cornwall with their bryophytes; and peat diggings in Norfolk with their rare damsel flies.[26]

An example in Scotland is Whitlaw Moss in Berwickshire, a National Nature Reserve, part of which was an old marl pit, dug six to eight feet below the surface in the eighteenth century once the top cover of peat had itself been removed. It is, nevertheless, home to a whole raft of *Red Data Book* plants and invertebrates. Sites like these do not exactly need habitat restoration, for that would be restoration to the old brutalities of extraction, but they do need management, and management needs understanding. The Nature Conservancy Council wisely, and unusually, commissioned a historian to write up all that is known about how the reeds, timber, fuel and fertiliser of the moss were exploited in the past. But they did not publish the report.[27]

---

[24] A. Holmes, 'Integrating history and ecology to sustain a living landscape: applying oral history narratives in interdisciplinary research', *Society for Landscape Studies Newsletter*, Winter 2007/8.

[25] See, for example, Peter Harvey, 'The East Thames Corridor: a nationally important invertebrate fauna under threat', *British Wildlife*, 12 (2000), pp. 91–8.

[26] Peter Middleton, 'The wildlife significance of a former colliery site in Yorkshire', *British Wildlife*, 11 (2000), pp. 333–49; Jeff Lunn, 'Wildlife and mining in the Yorkshire coalfield', *British Wildlife*, 12 (2001), pp. 318–26; Mark Crick, Pete Kirby and Tom Langton, 'Knottholes: the wildlife of Peterborough's claypits', *British Wildlife*, 16 (2006), pp. 413–21; Adrian Spalding, 'The nature-conservation value of abandoned metalliferous mine sites in Cornwall', *British Wildlife*, 16 (2005), pp. 175–83; Bob Gibbons, 'Reserve focus: Thompson Common Nature Reserve, Norfolk', *British Wildlife*, 15 (2004), pp. 240–3. See also Jonathan Briggs, 'The saga of the Montgomery Canal', *British Wildlife*, 17 (2006), pp. 401–10.

[27] Michael Robson, 'A history of the Whitlaw Mosses', typescript prepared for the Nature Conservancy Council, now in the Borders office of Scottish Natural Heritage.

There are two extensive habitats in upland Great Britain where the need for management and restoration is particularly clear, both of which call for deep understanding of their cultural past – heather moorland and raised bogs. Heather moorland was present in Britain on a small scale in the Mesolithic, although there is little evidence at present from palynology to suggest in the uplands any close equivalent to Vera's lowland savannahs. The great expanses of moor with which we are familiar today in Scotland and the north of England have been maintained as cultural landscapes free of the natural succession of tree cover only by, in the first instance, farmers burning and grazing the heather for millennia, and more recently by the managers of sporting estates burning and grazing in a subtly different way.[28] Grouse moors are critical habitat for a whole range of species from the golden plover to the *Bombus jonellus* bee. They are also the sites of great controversy between landowners and conservationists concerning how far the maintenance of large numbers of predators, such as hen harriers and peregrines, is compatible with large numbers of red grouse, on which the economic viability of the moors and therefore their continued management depends. If nature was to take its course, or the rewilders with their eye on 6,000 square kilometres of potential montane woodland to succeed, both the game and the raptors would be lost, as well as all the incidental species. Even as it is, heather moorland is in steep decline throughout western Europe, and in northern Britain in particular there is a responsibility to maintain it both as biodiversity resource and cultural icon.

If it is to be treasured, it is a historical question as to how this habitat was managed in the past. It may be enough to say in this and similar situations that we will maintain 'traditional' management, which essentially means what the more old-fashioned managers are doing now. But it might be even better to try to find out (if it is possible) what had been the earlier management options, for example, pre-gun, pre-trap, pre-flame-thrower, with different types and sizes of animal, and see if you like the biodiversity of that time more than that of the present day. Both are a product of culture, of particular systems of management, and there is a question of choice (constrained of course by possibility, not least economic possibility) as to what biodiversity scene you may wish to have at the end of the day.

---

[28] R. A. Dodgshon and G. A. Olsson, 'Heather moorland in the Scottish Highlands: the history of a cultural landscape, 1600–1880', *Journal of Historical Geography*, 32 (2006), pp. 21–37; C. M. Gimingham, 'Heaths and moorland: an overview of ecological change', in D. B. A. Thompson, Alison J. Hester and Michael B. Usher, *Heaths and Moorland: Cultural Landscapes* (Edinburgh, 1995), pp. 9–19; I. G. Simmons, *The Moorlands of England and Wales: an Environmental History, 8000 BC–AD 2000* (Edinburgh, 2003).

Raised bogs also show the paramount need for knowledge of past management. It is unclear how far our peat bogs are truly natural, but their formation accelerated with climatic change around the time of the onset of the Bronze Age, with more rain and wind. Human clearance of trees and cultivation may then, and at other times, have accelerated their growth. Many bogs overlie prehistoric field systems, but others are at least 9,000 years old and predate agriculture. All were used for millennia for the provision of food, pasture, fuel and building material, and it was this use that kept them in their distinctive form, especially the regular surface burning and grazing which prevented the resumption of tree growth and encouraged the continuous formation of the peat.

When, at various times over the last half century, such bogs came under the care of conservation bodies, they were ignorant of all this. Misunderstanding what had kept them in being so long, they enclosed them against stock and banned the use of fire. The almost uniform result has been that bogs in the care of conservation bodies have become progressively more overgrown with wood, while those outside conservation care have fallen victim to commercial peat extraction (not, as in Ireland, to fuel power stations, but in Britain usually for garden centre material). Frequently in the twentieth century, there was a judgement of Solomon, but in this case the division of the baby was carried out. Part of a great raised bog was used for commercial extraction and part was surrendered to conservation. Babies do not thrive if cut in two. Extraction lowered the water table, to accelerate further the afforestation of the conservation sections. The reaction has been to try to stop or to limit commercial use, with lesser or greater success in the case of two extremely important sites, Thorne Moors in Yorkshire and Flanders Moss in Stirlingshire.

But what do you do about the trees once you have them? You might of course welcome them in the spirit of accepting a change that has probably happened before over the millennia, but most conservationists felt that tree cover would change the biodiversity they had inherited, and decided to extirpate them. Unable to believe that grazing and burning would provide a solution, the conservation managers often tried to destroy the wood by hand and spray, a battle against birch seed that surely will never succeed for long. Some managers, hearing that burning and grazing might be the answer, rushed in without asking what kind of burning and grazing had been practised in the past. If they used a hot slow fire travelling against the wind towards unburned fuel, this could have a disastrous effect on the *Sphagnum* moss of which the moss is composed, and if they used sheep that were unused to grazing on bog land, some of the poor beasts would drown. Had they learned from past management how to do it correctly – with fires set to travel rapidly downwind and beasts with more inherited local knowledge – they would have done better. A peat bog, like a grouse moor, is a cultural landscape that needs exact information about its past use.[29]

ON NATURE AND CULTURE

The terms 'natural' and 'cultural' have appeared often in this paper because conservation scientists usually regard them as important distinctions. Generally speaking, the more something is 'natural', or wild, the more it is to be treasured, the more it is authentic.[30] If it is 'cultural', or anthropogenic, the less it is admired. Thus, with alien species, the natives, or non-aliens, have top conservation priority because their broad distribution can be taken to be natural. How this figures with swallows, collared doves, house sparrows or other such commensal species is an open question: they are assumed to have arrived on their own wings, naturally, but their habitat is totally determined by human culture. Archaeophytes, or aliens that arrived through man's cultural activity and applied for a residency permit more than 500 years ago, can be accepted but are apologised for. Neophytes, aliens that have had an assisted passage since 1500, are beyond the pale, and never protected. Nevertheless, some of these may be of both cultural value and international conservation concern, like Lady Amherst's pheasant in Bedfordshire, just as much as those threatened archaeophyte weeds.

But supposing a historian were to argue that maintaining the distinction between nature and culture is itself a cultural construct rather than a scientific fact? If man is part of nature then his culture must be part of nature too, and the distinction between 'natural' and 'wild' on the one hand, and 'cultural' and 'anthropogenic' on the other, starts to change. And in what sense is man not part of nature? That it could be otherwise is part of a pre-Darwinian intellectual world, based on the Christian notions of the Divine separation of man from nature, of man's divinely sanctioned command over nature, of the corruption of man, and of the primal innocence of the rest of nature (this last a Romantic rather than a Christian belief). Then, all importantly, the fission between the human and the natural was confirmed independent of religion or sentiment (so untainted by either) by the

---

[29] R. Hingley and H. A. P. Ingram, 'History as an aid to understanding peat bogs', in T. C. Smout (ed.), *Understanding the Historical Landscape in its Environmental Setting* (Dalkeith, 2002), pp. 60–88. See also L. Parkyn, R. E. Stoneman and H. A. P. Ingram (eds), *Conserving Peatlands* (Wallingford, 1996); J. H. Tallis, R. Meade and P. D. Hulme, *Blanket Mire Degradation: Causes, Consequences and Challenges* (Aberdeen, 1997).

[30] But note the explicit recognition of the cultural origins of heather moorland in Thompson, Hester and Usher (eds), *Heaths and Moorland*.

rationalism of the Enlightenment. It was established as a category different from culture or nurture and in opposition to them, and secular modern science came also to be 'built on a philosophical platform that makes nature separate from human society'.[31]

Yet Darwinian science blows this apart. Darwin placed man as part of the tree of evolution, a twig no higher than any other, the outcome of a natural process that might have no creator at all. Huxley, no doubt to make evolution more palatable to his Victorian consumers, still intellectually fed by centuries of assuming the primacy and separateness of man, placed man as the top of the tree, the final triumphant bud at the end of the leading shoot to which all evolution pointed. This was a much less radical conclusion than Darwin's and much less disturbing to the hubris of contemporary Victorian and modern ways of thought.[32] The twentieth century, including conservation science, remained broadly content with Huxley's way of singling out man from the rest.

If we take a more thoroughly Darwinian view, man's actions appear scientifically as natural as a limpet's. But if our actions are entirely natural, are they then entirely excusable? Is to place a supermarket on a wildlife site as valid as saving the whale? How, then do we avoid ethical anarchy?

Here we should set aside science and consider the reason why culture has been separated from nature in common speech. We speak of nature because the otherness of the non-human is so impressive and so precious to us. In Adam's words:

> It is always there, enduring our attempts to direct it by science and technology and to tame it with our cultural constructions ... People need that sense of the wild, and conservation needs that contact between individual people and other species and landscapes. It needs the energy that flames out from the recognition of nature.[33]

We may to a degree manipulate, but in the last resort we cannot control, the laws of the universe. Through taxonomy and biology, we may study but never fully comprehend the lives of other species. We cannot place a thermostat on the sun or enter the mind of a swallow. Mystery is compounded by the compelling beauty of the other, forever real, forever in the eye of the beholder, seen at sunrise, in swallow-flight, and in every other contemplation of natural phenomena.

[31] Adams, *Future Nature*, p. 104.
[32] Peter J. Bowler, 'Darwinism and Victorian values: threat or opportunity', in T. C. Smout (ed.), *Victorian Values, Proceedings of the British Academy*, 78 (1992), pp. 129–48.
[33] Adams, *Future Nature*, p. 105.

So if we do choose to refuse permission for the supermarket or fight to save the whale, it is not actually because of the scientific interest of butterflies and whales, though in our modern embarrassment about the unscientific we may pretend it is. It is actually because we value the otherness of other things.

The only way to avoid ethical anarchy is to return to a spiritual view of stewardship through recognising that, however it came about, man has exceptional powers both of command and of moral choice, and that this gives us a duty to protect the other – the rest of the natural world. Duty and self-interest turn out to be identical because culture needs nature: 'I am part of nature and I damage myself when I damage it.'[34]

In acting as stewards, however, we should think much more holistically and historically about what it is we should protect. Whether you are a rewilder or a regardener or something between, there is no logical reason whatever not to celebrate, preserve and respect what came about, or was favoured, by human cultural activity. I would go further, and argue that nature conservation will succeed much better in Britain when the assemblages of habitats and species are seen as part of our historical heritage, shaped by our culture, and valued as such. It will be better accepted when it becomes well integrated with historical interpretation of each site. Every ancient wood has its history and deserves interpretation in a booklet or on-line for the local community and the visitor: the sheets of bluebells are an ancestral gift from generations of woodmen departed. The partnership dedicated to returning the fens to East Anglia will surely not wish to ignore the history of the hunter, the thatcher, the drainer and the farmer, or omit to explain it alongside the history of the biodiversity, because the relationship between the two tells so much, and the explanation will gather public interest and support to its side.

'Rewilding', 'conservation gardening' and 'landscape-scale change' each exist in rich variety. They represent a spectrum of different approaches, all valid in context, if not all equally valid in every context. They are all cultural actions undertaken for our own satisfaction, and their best outcomes are great manifestations of the human spirit as much as manifestations of nature. They are what the natural world can no longer do unaided. We should not be apologetic about that, but we should celebrate and explain the intertwining of man with the rest of Creation, or, if you prefer, with the rest of evolved nature.[35]

---

[34] Mark Tulley, 'Something understood', broadcast on BBC Radio 4, 14 May 2006.
[35] An aspect of British nature conservation not touched on here is the reintroduction of extinct native species, which invariably involves archaeological and historical research on past status. See, as an outstanding example, B. Coles, *Beavers in Britain's Past* (Oxford, 2006).

Fig. 13 The Old Shoe and the Sea. In 2007, the Scottish Coastal Archaeology and Problem of Erosion Trust (SCAPE) organised a competition for the public to send in their images of the coastal environment. The winner in the under-18 section was Joe Niven, aged 8, from Unst, Shetland. Photographer: Joe Niven. Reproduced with permission.

CHAPTER 13

Environmental Consciousness*

In the context of nature conservation and environmental history, what exactly do we mean by 'environmental consciousness'? One thing is certain, that unless we have a love and respect for the natural world that is widely diffused both among our leaders and the rest of us, conserving nature will not occur except as an accidental consequence of humanity's other preoccupations.

The Hebrides, perhaps especially Morton Boyd's beloved Tiree, have a precious biodiversity, much of it the result of past and present use by the Hebrideans. But this is an accident of economic history, not the fruit of wise ideology. The rushy fields and the machairs, with their riot of flowers, their corncrakes and corn buntings, and their lovely bee, *Bombus distinguendus*, survived because the mechanical and chemical revolution in twentieth-century farming was not entirely appropriate for the specialised nature of the islands (exposed, remote and marginal), even if the crofters had had the capital to take full advantage of it. Tiree may be a naturalist's paradise and Suffolk a barley baron's desert, but that is not because the Gael is more in touch with nature than the Saxon. Given the chance to make money and save toil, the Scottish crofter is as quick on the uptake as the English farmer, and with as little thought to nature ('which can look after itself'), crowding the hill with subsidised sheep or switching from traditional hay stooks to black-bag silage. To say this is not to attempt to deny the right of a crofter to a decent living, or to say for a moment that there are no environmentally conscious Hebrideans. But the difference between the way the land is used in the Hebrides and in Suffolk is not due to some difference in outlook but to a great difference in economic opportunity. Whatever it may be, environmental consciousness is not an inborn ethnic virtue.

So what is it? It is immensely old, but it has not always expressed itself in the same way. It is a set of feelings of love and respect towards the natural, non-human world, expressed in a multiplicity of forms, some ancient with

---

* This is a slightly revised version of a tribute essay to J. Morton Boyd, the Nature Conservancy Council's director in Scotland in the 1970s, published in a memorial volume, R. Crofts and I. Boyd (eds), *Conserving Nature: Scotland and the Wider World* (Edinburgh, 2005), pp. 102–17. Thanks to Hugh Andrew of Birlinn for permission to reproduce.

roots so deep that we can hardly trace them, some recent in international treaties. We can expect it to change in the future as it has in the past. Because it has so many shades of meaning, seldom coherent or fully logical in their expression, it may appear to the cynical to be inconsequential. But its very universality is its power. It may range in its expression from my neighbour's wish to tell his friends about seeing a dragonfly or an albino swallow, to the gut feeling of a walker in the countryside that 'I love the outdoors', to a fisherman's glee at the tug on the line. It may be a carefully expressed philosophical position by a self-declared 'green', consciously espousing Aldo Leopold's land ethic or Arne Næss's deep ecology. For every one of the disciples of Leopold or Næss, there will be tens of thousands of ordinary people who just love nature.

At its fullest, environmental consciousness might be described as a respect for species not our own and a love for the beauty of all natural things. It encompasses a sense that living things comprise a web governed by nature's ecological and physical laws, and an awareness that what we do to modify the operations of any of these laws may impact on other species, perhaps all other species, including ourselves. It does not deny a human 'right' to 'exploit' nature for use or pleasure but, as most commonly articulated now, it includes a sense that such exploitation should be 'sustainable', capable of being continued from generation to generation without diminishing the delightfulness of the natural world or sacrificing the material and spiritual options of those who will follow us. Rather few of the tens of thousands who just love nature could articulate a feeling about sustainability. Yet any environmental consciousness that does not include a sense of the need for sustainability will throw too feeble a light to be a guide for the twenty-first century. This is the main challenge for environmental education at this cusp of planetary history.

Perhaps some consciousness embodying a sustainability ethic did once exist as a primitive virtue in our Mesolithic ancestors, who for the first 5,000 years of settlement history were the only inhabitants of Scotland: some anthropologists believe that they find it still in hunter-gatherers at a similar stage of development among the Arctic Inuit, in the Amazon basin or in New Guinea. But the historian and the archaeologist studying these islands cannot make assumptions by analogy, however tempting, about the ideology of those who lived too long ago and with too ephemeral a material culture to leave any trace of their belief systems. The same is broadly true of their Neolithic, Bronze Age and Iron Age successors, though by the time the last named begin to meld into the recognisable Celtic culture of the Picts, one can see from the sculptured portrayal of birds, fish and beasts a sophisticated sense of the animal kingdom. The beautiful nature poetry of

the ninth- and tenth-century Irish monks is a still more-eloquent testimony
to their feeling for the non-human world:

> May-time, fair season, perfect its aspect; blackbirds sing
> a full song, if there be a scanty beam of day.
> Summer brings low the little stream, the swift herd makes
> for the water, the long hair of the heather spreads out, the
> Weak cotton grass flourishes.
> The harp of the wood plays melody, its music brings
> Perfect peace; colour has settled on every hill, haze
> on the lake of full water.

The touch of such poems, observes James Hunter, 'is sure and delicate. They
indulge neither in elaborate description nor in introspective moralising.'[1]
And while agreeing that European poetry elsewhere had no equal to this at
so early a date, Hunter remarks that at a similar time Buddhist and Shinto
writers were equally speaking of the spiritual delight of the green mountains
and the spot where 'peach blossom follows the moving water'. So, over a
thousand years ago, poetic expression of love and respect for nature, though
not necessarily universal, found worldwide expression.

With the establishment, however, of the full rigour of the Roman-Judaic
Christian tradition – in Scotland from the eleventh century – the theologi-
cal emphasis was on God the Father who made Heaven and Earth, and then
made man in his likeness and on the fifth day of creation enjoined him be
fruitful and multiply, to fill and subdue the earth and to have dominion
over all living things. As John Lister-Kaye observed once, we have certainly
ticked that box. The Genesis myth has often been interpreted, even by
modern Protestant Christians, as giving man license to do as he wished with
the natural world. I well recall a forester from one of the companies most
closely involved with the attempt to plant up the Caithness Flows saying at
a conference that as a Christian he did not believe in conservation, as the
world was a passing thing and man had a duty to his Maker to exploit it to
the uttermost. Ronald Reagan's Secretary of the Interior, James Watt, publicly espoused a similar creed when he was in charge of the US Parks and
Wildlife Service, further maintaining that there was no point in environmental protection as God was preparing his apocalypse.[2] Throughout this
essay, I use 'man' simply as a translation for the species name *Homo sapiens*,

---

[1] J. Hunter, *On the Other Side of Sorrow. Nature and People in the Scottish Highlands*
(Edinburgh, 1995), p. 47.
[2] J. R. McNeill, *Something New Under the Sun: an Environmental History of the Twentieth
Century* (New York, 2000), p. 352.

though in most cases such as these, involving the crass exploitation of nature the gender connotation would be especially appropriate.

But the Roman-Judaic Christian tradition, in this as in most things, is ambivalent: alongside the dominion claim there also runs an ethos of stewardship. Against Adam, the Lord of Creation, we can set Noah, the first conservationist, with a famous interest in biodiversity. The nature-mysticism of St Francis was a countervailing weight to St Augustine, whose teachings were taken as warning that those who admired mountains, rivers and seas were admiring what was earthly and imperfect and distracting themselves from the proper contemplation of God, who was eternal and perfect. That Augustine's abstract contemplation of the unearthly did not entirely prevail can also be most clearly seen in the work of late medieval artists, who carved realistic foliage on Gothic columns and memorably portrayed a wren in pursuit of a spider in the stained glass of York Minster.

Nevertheless, the Middle Ages in Europe left us with a man-centred view of the world about us. If the heavens showed forth the glory of God, living creatures on earth were seen only in relation to humanity.[3] The ox and the ass were useful; the rat and the flea were a demonstration of God's displeasure, a punishment for fallen man; others were there to symbolise moral qualities. 'Go to the ant, thou sluggard, and consider his ways', was the admonition of the Jewish psalmist still used by the seventeenth-century preacher; the brave lion, the cruel wolf, the busy bee were part of the classical world of Aesop's fables; the pelican who embodied self-sacrifice by supposedly feeding her young from blood from her own breast was a monkish fantasy.

Nature, in other words, taught lessons. King James VI of Scotland was a representative intellectual of the late Renaissance. In a well-turned sonnet, he listed the wonders of nature:

> The azured vault, the cristall circles bright
> The gleaming firie torches poudered thair
> The sadd and bearded fires, the monsters faire
> The prodiges appearing in the aire
> The rearding thunders and the blustering windes
> The foules in hewe, in shape and nature rare
> The prettie notts that wing'd musicians findes
> In earthe, the sauorie flowers, the metall'd mindes
> The wholesome herbes, the hautie pleasant trees

---

[3] K. Thomas, *Man and the Natural World: Changing Attitudes in England 1500–1800* (London, 1983).

> The silver streames, the beasts of sundrie kindes
> The bounded roares, and fishes of the seas
> All these for teaching man the Lord did frame
> To honoure him whose glorie shines in thame.[4]

There are continuities in our attitudes today with this world-view, especially the idea of the whole of nature as marvellous. But there are also discontinuities and transformations: the monsters and prodigies could hardly withstand the early scientific revolution, the Baconian insistence on gathering empirical facts and the Newtonian obsession with discovering universal laws. But the early scientists themselves were hardly materialists: they developed medieval ideas into a notion that the nature of the Creator could be revealed by a study of his creation. The early English naturalists such as William Turner in the sixteenth century and John Ray in the seventeenth regarded their studies as an act of piety, and ultimately the idea coalesced into the theory that 'natural religion' (the study of nature and nature's laws) could supplement 'revealed religion' as a window into the mind of God.[5] This was a particularly satisfying idea to the eighteenth century, exhausted by the bloody conflicts of the seventeenth century that had hinged round subjective interpretations of God's will as revealed in scripture.

It is fascinating to see the transition of the world of the fable to the world of biology, and to consider how eighteenth-century science (surely, as modern science) reflects also the social ideas of the day. Benjamin Stillingfleet expressed a notion both remarkably close to the modern concept of 'the web of life' and heavily infused with contemporary theory of 'a Great Chain of Being' that kept society stable:

> . . . each moss,
> Each shell, each crawling insect holds a rank
> Important in the plan of Him, who fram'd
> This scale of being: holds a rank, which lost
> Would break the chain and leave behind a gap
> Which Nature's self would rue.

It is well known that seventeenth-century Lady Glanville was considered mad for collecting insects, and when her heirs disputed her will John Ray

---

[4] *The poems of James VI of Scotland*, ed. J. Craigie, Scottish Text Society, Third Ser., 26 (1952), vol. 2, p. 99.

[5] C. E. Raven, *English Naturalists from Neckham to Ray* (Cambridge, 1947); C. E. Raven, *John Ray Naturalist. His Life and Works* (Cambridge, 1950).

appeared to testify to her sanity. Stillingfleet (in the words of Thomas Pennant) 'attempted to destroy the false shame that attended the devotee to Ornithology, the chase of the Insect, the search after the Cockle, or the poring over the Grass'.[6] Within a generation, though, false shame and imputations of madness were removed by universal admiration for the labours of Gilbert White in England and Carl von Linné in Sweden, clergy or the son of clergy in their respective national established churches, the one devoted to exact field study and the other to systematic taxonomy.

By the nineteenth century, to be a natural historian with a collection of birds' eggs, butterflies, moths, seashells, pressed flowers and ferns had become a national pastime for many Britons of both sexes: guns, nets, vascula and killing bottles were complemented by microscopes, libraries and specimen cabinets with meticulously written labels.[7] The Victorian naturalists, both professional and amateur, made extraordinary strides in the taxonomic understanding and geographical distribution of the flora and fauna of these islands. All our modern understanding of biodiversity rests originally on their groundwork. Some of those responsible in Scotland are not forgotten, like William MacGillivray, author in 1837 of the five-volume *History of British Birds*, or J. A. Harvie-Brown, whose pioneering *Vertebrate Fauna* of Scotland's natural regions and studies of red squirrel and capercaillie, are lasting monuments of early natural history. Others are totally forgotten, like John Hood, the metal-turner of Dundee, discoverer of scores of species of rotifer new to science.[8]

What, of course, became displaced in the nineteenth century was the absolute confidence in what William Paley had called in 1802 'evidence of the existence and attributes of the deity collected from the appearances of nature' (the subtitle of his famous book on *Natural Theology*). Geology had first upset the applecart, with the replacement of biblical time by geological time. The subject became obsessive in Scotland with the work of James Hutton, Louis Agassiz and Charles Lyell. The fight to reconcile geological science with scripture without retreating into fundamentalism dumb and blind, reached its peak in the work of the stonemason, journalist and amateur geologist, Hugh Miller. In *The Old Red Sandstone* of 1841, he tried to reconcile 'the revealed truths of religion with the material proofs of

---

[6] Quoted in G. Taylor, 'Insect life in Britain', in W. J. Turner (ed.), *Nature in Britain* (London, 1983), pp. 260, 291.

[7] D. E. Allan, *The Naturalist in Britain. A Social History* (London, 1976); L. Barber, *The Heyday of Natural History, 1820–1870* (London, 1980).

[8] R. Ralph, *William MacGillivray* (London, 1993); B. McGowan and David Clugston, 'Pioneers in Scottish ornithology', in R. W. Forrester and I. J. Andrews (eds), *The Birds of Scotland* (Aberlady, 2007), vol. 1, pp. 97–106; A. D. Peacock, 'John Hood of Dundee: working-man naturalist', *Scots Magazine*, 31 (1939).

science'.[9] *The Testimony of the Rocks* (subtitled 'Geology in its bearings on the two theologies, natural and revealed'), and *The Footprints of the Creator*, were written in indignant response to Robert Chambers' *Vestiges of Creation* of 1844, that traced an early evolutionary principle in nature.

Then came Darwin. What Agassiz had called (in relation to Miller) 'the successful combination of Christian doctrines with pure scientific truth' seemed finally to be blown out of the water in 1859 by *The Origin of Species*. Evolutionary theory has been the ruling paradigm in biology since the late 1860s, ruling out a single act of creation of all the species on earth and undermining the need for a creator. The shackles of 'natural theology' were thrown off scientific thinking. Yet it was never Darwin's aim to disprove the existence of God, and some with faith can reconcile Darwinism with Christianity by postulating a creator who devised a system so ingenious that it would go on creating itself. More importantly, there are few who study the natural world either as amateurs or as professionals who do not stand in awe of its beauty and intricacy, just as Darwin himself did. A great many thereby find in it a spiritual as well as a physical dimension. Morton Boyd was certainly among them, and differed from most of his colleagues mainly in the happy openness with which he admitted and enjoyed it. King James' notion that glory shines in nature is still part of our attitude to the environment.

One unexpected and radical thing that Darwin did, which is still not properly appreciated in popular conceptions of his work, was to dethrone man's supremacy. As P. J. Bowler has explained, evolution for Darwin was a tree with many branches but no leading shoot, and the branches had many twigs, which were modern species, each adapted, but no-one higher than another. In the hands of Thomas Huxley and others, however, 'Darwin's emphasis on the branching character of evolution was suppressed by implying that the tree of life had a central trunk defining the ladder of progress towards mankind'.[10] So western thought merely swapped the idea of man made in the image of God (but fallen and imperfect) for the much more vain-glorious one of man made by evolutionary law as nature's most perfect expression – the winner, in fact, in all circumstances. That is one root of our pathetic modern hubris. If Darwin had been properly understood, our value would seem to be the same as that of any other species and our self-assurance totally unjustified.

But most people surely do not relate to the environment through any knowledge of biology but through ancient forms of visceral feeling. The

---

[9] C. D. Waterston, 'An awakening interest in geology', in L. Borley (ed.), *Hugh Miller in Context* (Cromarty, 2002), p. 89.

[10] P. J. Bowler, 'Darwinism and Victorian values: threat or opportunity', in T. C. Smout (ed.), *Victorian Values* (Oxford, 1992), p. 138.

hunting instinct is one of the most powerful of these – its artistic expression can be traced back to the Palaeolithic caves of the Mediterranean world, to be found again in Pictish sculpture and medieval tapestry. The medieval and early modern view of the world in Europe did not imply that hunters had any respect for their quarry, and in an age when Descartes could maintain that creatures other than man were mere machines, a concept of animal rights was not likely to arise. King James IV used to combine pilgrimages to the Isle of May with taking the local lairds and canons of Pittenweem to shoot at the nesting seabirds with his culverins. Two hundred years later, an account of Handa in 1726 said that the island abounded with breeding seabirds, and that 'going there in a calm day with a boat and some fowling pieces is an agreeable diversion'. A century later than that, when sensibilities towards other species had indeed so far developed as to make this unacceptable, it was day trips from Scarborough to shoot at the nesting birds of Bempton Cliffs that helped to trigger early moves towards bird protection with acts in 1869 and 1880.[11]

The legislation did not come a moment too soon, as by then even the gannets on the famous colony on the Bass Rock were in decline: no longer much fancied as food, in the age of cheap guns and cartridges it was regarded as fun to use the birds for target practice. The Prince of Wales joined in, arriving by special train in 1859, and travelling out to the Bass in a yacht accompanied by a gunboat, HMS *Louisa*, he 'enjoyed some shooting and brought down a number of Gannets'.[12] The story of our treatment of other living things for our amusement is not generally edifying, but times improve. In 2000, another and more enlightened Prince of Wales sailed round the Bass armed only with his binoculars, after he had opened the Scottish Seabird Centre at North Berwick.

Yet hunting was an important way of encountering nature – for some it still is – and it had a transforming capability that was well captured in 1714 by Arthur Stringer, keeper to Lord Conway of Portmore in Ulster, in a book partly addressed to his employer's friend, the Earl of Antrim.[13] He described him 'in all the heroic airs of that diversion' standing in his saddle: 'Bellowing to your hounds, your wig wafted by the winds, your eyes sparkling with gladsome joy, and your whole mein expanded, as it were opened out, thrown abroad to the exulting extasy.' Wordsworth among his daffodils was not more transformed. Stringer later speaks with the hearty tones

[11] N. Macdougall, *James IV* (Edinburgh, 1989), p. 198; A. Mitchell (ed.), *Geographical Collections Relating to Scotland* (Edinburgh, 1906), vol. 1, p. 197; J. Sheail, *Nature in Trust: the History of Nature Conservation in Britain* (Glasgow, 1976), pp. 4, 23–4.

[12] W. M. Ferrier, *The North Berwick Story* (North Berwick, 1980), p. 68.

[13] A. Stringer, *The Experienced Huntsman*, ed. J. Fairley (Belfast, 1977).

expected from a twentieth-century proponent of youth hostels extolling the fitness benefits of an outdoor life:

> How pleasant and healthful it is to suck in the sweet fragrant air that the woods and fields afford in the morning early, which clears the lungs and pipes and purifies the blood, while the slothful lie in bed to nourish or foster their sloth and lust, coughing and spitting.

Centuries later, this is entirely recognisable by the hunting fraternity in all its forms. Stringer can be improbably matched by the Danish novelist Isak Dinesen (Karen Bliksen) writing in 1960 of her absorbed pursuit of lions and elephants in mid-twentieth-century Kenya, but now a note of respect for the quarry is added:

> Hunting is ever a love-affair. The hunter is in love with the game, real hunters are true animal lovers. But during the hours of the hunt itself he is more than that, he is infatuated with the head of game which he follows and means to make his own; nothing much besides it exists in the world.[14]

One thing that has changed is that the hunt is now pursued in all sorts of ways that Stringer, at least, would have found baffling. The obsession with the chase is found not only in the membership of the Countryside Alliance but in the twitchers who gather in their hundreds on the Scilly Isles to spot the October bird rarities from America and Siberia, or who drive through the night, 'infatuated with the head of game', from London to Nigg to see a Blyth's reed warbler: 'nothing much besides it exists to him in the world'. In describing, as a teenager, his first encounter with a red-backed strike tracked down in the New Forest, Chris Packham unconsciously echoes Isak Dinesen: 'I remember my whole world condensing round this one little bird. Nothing stirred or was even alive outside my heartbeat and the bird.'[15] The Victorian naturalists who mapped our biodiversity were also yielding to their hunting instinct, only too literally in the case of those egg collectors who did so much harm to the Scottish birds. But, after reading his remarkable books, we cannot doubt the obsession of Charles St John who shot the last ospreys in Sutherland, or deny that he experienced the transforming power of the natural world while he was doing it.

[14] I. Dinesen, *Shadows on the Grass* (London, 1960).
[15] C. Packman, 'The blasted heath', *Sanctuary: the Ministry of Defence Conservation Magazine*, 32 (2001), p. 42.

The Victorians also added to our experience of nature the cult of athleticism. The countryside also began to be seen as an obstacle course in which people could pit their muscle and endurance by climbing, walking and skiing in lonely places – and now also by canoeing, mountain-biking, windsurfing and hang-gliding, or even by the anti-social pastimes of jet-skiing, motorboating and motorcycle scrambling.

The shortcoming of the hunting and athletic ways of experiencing nature is that man is unquestionably put first. For all the claimed respect for the quarry, the codes of conduct and the partnership agreements, it is the pleasure and sense of achievement by the individual in overcoming the quarry or conquering the obstacle that is the ruling force. That is not to deny hunters the credit for doing much good – the Game Conservancy and The British Association for Shooting and Conservation can be a force for habitat conservation that other environmentalists underestimate or overlook. But it is only when something else is added to hunting and the athletic ways of experiencing nature that they become forms of environmental consciousness that could lead to an ethic of sustainability. That missing factor is a spiritual sense of elation and peace in the presence of nature. It is difficult to talk about because it is hard to define, but also because a materialist age embarrassed to speak of the spirit is losing the language to do so. But its fruit is an understanding that nature after all is not for people.

Nature is for itself. In the words of Morton Boyd's great mentor, Frank Fraser Darling:

> Wildlife does not exist for man's delection. Man may find it beautiful, edifying, amusing, useful and all the rest of it, but that is not why it is there, nor is that a good enough reason for allowing it to remain. Let us give beast and bird and flower the place to live *in its own right*, unless pest or obnoxious parasite. If we can admit that right and desire to accord it, we are well on the way to taking a holistic view of wild life in the economy of the nation.[16]

Or, taking a step further, here are the words of a modern Quaker, Audrey Urry:

> All species and the Earth itself have interdependent roles within Creation. Humankind is not *the* species, to whom all others are subservient, but one among many. All parts, all issues are inextricably intertwined. Indeed the web of Creation could be described as of three-ply

---

[16] F. F. Darling, *The Natural History of the Highlands and Islands* (London, 1947), p. 263.

thread: wherever we touch it we affect justice and peace and the health of all everywhere.[17]

Those who come to see these insights are empowered and impelled to use nature with the gentleness that is implied by the concept of sustainability. If we do not have a sense that nature is for itself, then we see it as primarily an article of human consumption. The twentieth century's history of consumption does not suggest that we in the twenty-first century would know when to stop.

A kind of transcendent delight in nature we have already traced as an element in human culture at least as far back as the Irish and Chinese poets of the sixth to ninth centuries. It is worth backtracking in the chronological development of our argument to bring this out. One can sense such a delight rising in Petrach in the fourteenth century as he climbed the hills of Provence and looked back into Italy, but he tells us that he fought it back by thinking about St Augustine's warnings against admiring earthly splendour. In Virgil and the Roman poets, delight concentrates in a deeply felt pleasure in a harmonious countryside in which man and nature are closely related in good husbandry and the seasonal rhythms of the natural world. So there is something very old and also various in the expressions of delight in the environment that encompasses us.

In Renaissance Scotland, pleasure in nature suffused much of the poetry of the sixteenth-century court circles, with Virgil the obvious model. In the prose writings of the seventeenth-century topographers, trying as accurately and soberly as possible to describe Scotland for the first time, the tone was also classical, concerned with rural plenitude even in the remotest counties of the Highlands. Thus the forests and chases of Sutherland were 'very profitable for feeding of bestial and delectable for hunting', the straths 'well manured and inhabited, replenished with woods, grass, corns, cattle and deer, both pleasant and profitable', and Dornoch had 'the fairest and largest links or green fields of any part of Scotland, fit for archery, golfing, riding and all other exercise. They do surpass the fields of Montrose or St Andrews'.[18]

It is customary, and with good reason, to trace much of our modern sensitivity about the environment to the Romantics. What they particularly gave us, which had been missing in the Renaissance, was an appreciation of the drama and detail of wild nature – an excitement about what had not been tamed by man as opposed to the Virgilian concept of nature under control.

---

[17] Society of Friends, *Quaker Faith and Practice* (1995).
[18] A. Mitchell (ed.), *Geographical Collections Relating to Scotland made by Walter Macfarlane* (Scottish History Society, Edinburgh, 1906) vol. 3, pp. 99–100, 103.

Many of the writers who framed this view came either from Scotland or from northern England, Wales or Ireland, in landscapes that south-eastern-ers found intimidating.

No one was more important in redirecting the vision of nature in the poetry of the British Isles than the young Roxburghshire poet James Thomson, who arrived in London in 1725 at the age of twenty-five, with his first version of his descriptive poem, *Winter*, part of what was to become the *Seasons*. He celebrated nature's 'majestic works' in snowstorms and floods – 'With what a pleasing Dread they swell the Soul' – and attended to the detail of insects pouring forth:

> . . . From every Chink
> And secret Corner, where they slept away
> The wintry storms; or rising from their Tombs
> To higher Life: by Myriads, forth at once
> Swarming they pour: of all the vary'd Hues
> Their Beauty-bearing Parent can disclose.
> Ten thousand Forms! Ten thousand different Tribes!

Thomson had an extraordinary liberating impact on British literature, among Gaelic poets as well as on those writing in English or Scots: they appeared now to be able to look nature in the eye without an overload of classical allusion or metaphor. In the seventeenth century, the poets of Gaeldom used nature as a device in the 'pathetic fallacy' – the curlew wails because the laird is dead, and so on.[19] Now poets like Mac Mhaighstir Alastair and Duncan Ban Macintyre described nature by what they saw. This is Macintyre:

> In the rugged gully is a white-bellied salmon
> That cometh from the ocean of stormy wave
> Catching midges with lively vigour
> Unerringly, in his arched, bent beak
> As he leapeth grandly on raging torrent
> In his martial garb of the blue-grey back
> With his silvery flashes, with fins and speckles
> Scaly, red-spotted, white-tailed and sleek.

On the English side, the inspiration of Thomson helped to ignite the genius of Crabbe and Clare, on the Scottish side that of Burns and Tannahill,

---

[19] D. Thomson, *An Introduction to Gaelic Poetry* (1905).

united in the simplicity and directness with which they described the natural world. Several of them came from humble origins and their compositions – like those of Macintyre and Burns – were popular among the common people. They were not inventing an elite poesy as much as disclosing a view of nature that struck a popular chord.

The compositions of the high Romantics, above all Wordsworth and Scott, were much more cerebral, more moralising, not free of the poetic fallacy, and based on an elaborate aesthetic theory that had evolved among British intellectuals since Edmund Burke's *Essay on the Sublime* (1756). They carefully distinguished between the various emotions that 'swell the soul' – the beautiful, the sublime, the picturesque, the terrific. Walter Scott, for instance, on his journey round the northern and western Highlands, wrote of the different emotional characteristics of different sea caves: Smoo Cave in Sutherland, with its 'savage gloom' evoked 'terror', 'terrific'; Macalistair's Cave on Skye had a 'severe and chastened beauty'; Fingal's Cave on Staffa, combining grandeur with beauty, was an expression of 'sublimity'. What the Romantic landscape critic above all sought – in a way that seems quite new – was the wild: the untamed wilderness for its own sake. For Scott, the apotheosis of wilderness was Loch Coruisk on Skye. He describes in his diary how he was taken, from a coastal inlet boiling with salmon and attendant sea-birds, up 'a huddling and riotous brook' until they came to 'a most extraordinary scene', the loch surrounded by mountains, the shores composed of huge layers of naked granite mixed with bogs and heaps of sand and gravel where the torrents descended: 'vegetation there was little or none and the mountains rose so perpendicularly from the water's edge that Borrodale is a jest to them.' Even in the most stupendous scenery in the Hebrides, he could not resist the very human feeling that it was nice to be one up on Wordsworth.[20]

It was not just the detail of nature; it was the interaction of the human spirit with the tremendous forces of the wildest places that so deeply moved the Romantics. We now say that Scotland has no wilderness, meaning that it has no place unaffected by human hand, but all things are relative. We still seek this interaction in our own country and find its humbling and consoling power in places like Loch Coruisk.

Then there was the rise of the humanitarian movement towards animals, the shift away from the Cartesian view of the rest of creation as insensate machines. The Quaker, John Woolman, movingly and perfectly expressed it as early as 1772:

---

[20] *Journal of Sir Walter Scott* (Edinburgh, 1999).

I was early convinced in my mind that true religion consisted in an inward life wherein the heart doth love and reverence God and Creator and learns to exercise true justice and goodness not only toward all men but also toward the brute creation; that as the mind was moved on an inward principle to love God as an invisible, incomprehensible being, on the same principle it was moved to love him in all his manifestations of the visible world: that as by his breath the flame of life was kindled in all animal and sensitive creatures, to say we love God ... and at the same time exercise cruelty toward the least creature ... was a contradiction in itself.[21]

By 1900, then, we had inherited a complex set of perceptions about the natural world. Darwin had shown that man's biological place in it was not what Christians had supposed for almost two millennia. The hunting ethic was still at its peak with record bags on sporting estates, but had broadened out into recording and collecting across a whole range of biota. A new athletic enthusiasm had emerged for mountaineering, walking, skiing, boating and enjoying nature in many other novel ways. The Romantics had added a love for wilderness to the pleasures of closely observing natural things and the Virgilean sense of place and love for the agricultural life lived in harmony with nature. With the advent of the RSPCA as a popular and well-supported charity, many cruel sports and ways of mistreating animals had been banned by law in recognition of man's duty towards his fellow sensate beings, and the first bird protection legislation had been passed.

These forms of consciousness flourished greatly in the twentieth century. Biology gave birth to ecology, a vibrant science in mid-century. Substantial environmental and cultural bodies arose, often from nineteenth-century roots – the RSPB with its million members by the end of the 1990s; the National Trust and the National Trust for Scotland, bigger still; the Wildlife Trusts covering every division of Britain; the Ramblers – along with new societies – the World Wildlife Fund, Friends of the Earth, Greenpeace – presenting even by the 1970s a political voice to be reckoned with. Large tracts of land had been purchased and set aside for nature by the voluntary bodies. Government from 1949 had established national parks, national nature reserves and SSSIs, along with agencies dedicated to their care. Visits to the countryside, made accessible by car, snowballed: there were 22 million day visits a year to the Peak District National Park alone by the 1990s. Driving, walking, riding, climbing, twitching, sailing – the British

---

[21] *Quaker Faith.*

population's love of the countryside and the outdoors had burgeoned beyond all imagining.

But of course it brought with it problems of its own. Lake Windermere and Loch Lomond became very different places from those that Wordsworth and Scott had known, as the roaring motorboat and jet ski found their place. Sheltered English coasts were enveloped by marinas. Scottish mountaintops were transformed by ski tows, and before the century was out work had begun on carrying a funicular railway to the summit of Cairngorm. The industries that catered for outdoor recreation, in other words, were never curbed by any popular ethos of respect for nature. Then there were conflicts of a different kind, as when animal rights protestors demanded an end to cruel practices in the farming or pharmaceutical industries: there was a conflict here between regard for other sensate beings and the interest of the consumer in cheap food and safe drugs. Voluntary bodies such as the RSPB, dedicated to the preservation of other species, found that their priorities might impinge on the interests or livelihoods of those who lived on or near their reserves. So there was plenty of scope for confusion and conflict between people expressing different forms of environmental consciousness, or between people expressing some kind of environmental consciousness treading on the toes of economic or professional interests. Despite the growth of all the different ways of appreciating nature, there was no overriding ethos of respect for it.

An even greater cause for anxiety was the increase in the rate of exploitation of the world's natural resources. World energy use, standing at 400 million metric tons of oil equivalent in 1800, stood at 1,900 million by 1900 and 30,000 million in 1990: it grew fivefold in the nineteenth century and sixteenfold in the twentieth. World population growth was from 1.0 billion in 1800 to 1.6 billion in 1900 to 6.0 billion in 2000 – close to a fourfold growth in this century. In J. R. McNeill's words, it is

> an amazing development, an extreme departure from the patterns of the past – even though we tend to take our present experience for granted and regard modern rates of growth as normal. Bizarre events that last for more than a human lifetime are easy to misunderstand.[22]

In the course of the twentieth century, carbon dioxide emissions grew seventeenfold, sulphur dioxide emissions thirteenfold and lead emissions to the atmosphere about eightfold: small wonder that by its close, human activity was (in the opinion of most scientists) beginning to change the climate of

[22] McNeill, *Something New Under the Sun*, p. 9.

the biosphere itself, and in a way that would lead to an increase in extreme and catastrophic events. Despite international agreement at Kyoto and elsewhere, few commentators expect the situation to improve, since as affluence spreads from the developed world (where 10 per cent of the population use 80 per cent of the resources) to China, India, Mexico, Brazil and Indonesia, the rate of consumption and emission must increase.

On this uncertain situation, a ray of still more uncertain light is thrown from the biological sciences. Biology had spawned not merely ecology (in Britain a discipline with less visionary zeal in 2000 than it had had in the 1940s) but molecular biology, and (out of that) ultimately genetic engineering, the consequences of which are simply unguessable. To some soothsayers, it is the next technological fix that will save the human population on the planet; to others, it is the step too far that could lead to the extinction of life itself. We live in startling and utterly uncertain times.

So is nature for people? There is a sense in which the history of the twentieth century makes the affirmative a truism: man has taken nature and transformed both his numbers and his standard of living by making unlimited use of its resources. But if we are to transform mere headlong profligacy into sustainable use – the *sine qua non* of our continued existence as a global civilisation – we have to say that nature is for nature. That is to say, we have to go beyond utilitarianism to reverence. It does not suffice to say that we must have the rainforest because it contains genetic resources that could be of benefit to humanity. Within a generation, we will probably be able to make all kinds of genetic material at will in the laboratory. Will that then justify the destruction of the remaining forests? It does suffice to say that we must save the rainforest because it is an obscenity to cause the extinction of another species and a wickedness to destroy beauty, mystery and complexity that we can never recreate. 'Where was thou when I laid the foundations of the earth?' God asked Job, 'When the morning stars sang together, and all the sons of God shouted for joy? Or who shut up the sea with doors when it brake forth as if it had issued out of the womb?'

We must find within ourselves the humility born of true Darwinism to see that man is not the top of the tree, but one of countless twigs, and the spirituality of ancient religious traditions to respect the delicate mystery and overwhelming majesty of creation. If we cannot discover these qualities of humility and respect within humanity, the future of this spinning earth may be a silence to match that of her sister planets.

Select Bibliography

Adams, W. M. (1996), *Future Nature: a Vision for Conservation*, London.

Allen, D. E. (1976), *The Naturalist in Britain, a Social History*, London.

Anderson, M. L. (1967), *A History of Scottish Forestry*, Edinburgh.

Bangor-Jones, M. (2002), 'Native woodland management in Sutherland: the documentary evidence', *Scottish Woodland History Discussion Group Notes*, 7, pp. 1–5.

Blaxter, K. and Robertson, N. (1995), *From Dearth to Plenty: the Modern Revolution in Food Production*, Cambridge.

Botanical Society of Scotland (1996), Symposium on Environmental History of the Cairngorms, *Botanical Journal of Scotland*, 48.

Botanical Society of Scotland (2005), Symposium on Atlantic Oakwoods, *Botanical Journal of Scotland*, 57.

Bowler, P. J. (1990), *Charles Darwin: the Man and his Influence*, Cambridge.

Bowler, P. J. (1992), 'Darwinism and Victorian values: threat or opportunity', in T. C. Smout (ed.), *Victorian Values, Proceedings of the British Academy*, 78, pp. 129–48.

Boyd, J. M. (1986), *Fraser Darling's Islands*, Edinburgh.

Chambers, F. M. (ed.) (1993), *Climate Change and Human Impact on the Landscape*, London.

Cherry, G. (1975), *National Parks and Recreation in the Countryside: Environmental Planning 1939–1969*, HMSO.

Clapp, B. W. (1994), *An Environmental History of Britain since the Industrial Revolution*, London.

Coles, B. (2006), *Beavers in Britain's Past*, Oxford.

Cronon, W. (ed.) (1995), *Uncommon Ground: Rethinking the Human Place in Nature*, New York.

Darling, F. F. (1947), *Natural History in the Highlands and Islands*, London.

Darling, F. F. (1955), *West Highland Survey: an Essay in Human Ecology*, Oxford.

Davies, A. L. and Dickson, P. (2007), 'Reading the pastoral landscape; palynological and historical evidence for the impact of long-term grazing on Wether Hill, Ingram, Northumberland', *Landscape History*, 29, pp. 35–45.

Dodgshon, R. A. and Olssen, G. A. (2006), 'Heather moorland in the Scottish Highlands: the history of a cultural landscape, 1600–1880', *Journal of Historical Geography*, 32, pp. 21–37.

Elton, C. S. (2000), *The Ecology of Invasions by Animals and Plants*, new edn with foreword by D. Simberloff, Chicago.

Foster, S. and Smout, T. C. (eds) (1994), *The History of Soils and Field Systems*, Aberdeen.

Fox, A. D. (2004), 'Has Danish agriculture maintained farmland bird populations?', *Journal of Applied Ecology*, 41, pp. 427–39.

Grant, I. F. (1961), *Highland Folk Ways*, London.

Hall, M. (2005), *Earth Repair: a Transatlantic History of Environmental Restoration*, London.

Hingley, R. and Ingram, H. A. P. (2002), 'History as an aid to understanding peat bogs', in Smout, T. C. (ed.), *Understanding the Historical Landscape in its Environmental Setting*, Dalkeith, pp. 60–86.

History and Theory (2003), 'The nature of environmental history', Theme Issue 42.

Hodder, K. H., Bullock, J. M., Buckland, P. C. and Kirby, K. J. (2005), 'Large herbivores in the wildwood and modern naturalistic grazing systems, *English Nature Research Reports*, 648.

Holm, P., Smith, T. D. and Starkey, D. J. (eds) (2001), *The Exploited Seas: New Directions for Marine Environmental History*, St John's, Newfoundland.

Hudson, P. (1992), *Grouse in Space and Time*, Fordingbridge.

Hunter, J. (1973), 'Sheep and deer: Highland sheep farming 1850–1900', *Northern Scotland*, 1, pp. 199–222.

Hunter, J. (1995), *On the Other Side of Sorrow. Nature and People in the Scottish Highlands*, Edinburgh.

Johnston, E. B. (2001), 'People, peatlands and protected areas: case studies of conservation in Northern Scotland', unpublished University of Aberdeen Ph.D. thesis.

Kirby, K. J. and Watkins, C. (1998), *The Ecological History of European Forests*, Wallingford.

Kjærgaard, T. (1994), *The Danish Revolution 1500–1800: an Ecohistorical Interpretation*, Cambridge.

Lambert, R. A. (ed.) (1998), *Species History in Scotland: Introductions and Extinctions since the Ice Age*, Edinburgh.

Lambert, R. A. (2001), *Contested Mountains, Nature, Development and the Environment in the Cairngorms Region of Scotland, 1880–1990*, Cambridge.

Lindsay, J. M. (1974), 'The use of woodland in Argyllshire and Perthshire between 1650 and 1850', unpublished University of Edinburgh Ph.D. thesis.

Lindsay, J. M. (1975), 'Charcoal iron smelting and its fuel supply: the example of Lorn furnace, Argyllshire, 1753–1876', *Journal of Historical Geography*, 1, pp. 283–98.

Lindsay, J. M. (1980), 'The commercial use of woodland and coppice management', in M. L. Parry and T. R. Slater (eds), *The Making of the Scottish Countryside*, London, pp. 271–90.

Lovegrove, R. (2007), *Silent Fields: the Long Decline of a Nation's Wildlife*, Oxford.

Marren, P. (1994), *England's National Nature Reserves*, London.

Marren, P. (2002), *Nature Conservation: a Review of the Conservation of Wildlife in Britain 1950–2001*, London.

Marren, P. (2005), *The New Naturalists*, London.

Martin, J. (2000), *The Development of Modern Agriculture: British Farming since 1931*, Basingstoke.

McNeill, J. R. (2000), *Something New Under the Sun: an Environmental History of the Twentieth-century World*, New York.

Mei, X. (2007), 'From the history of the environment to environmental history: a personal understanding of environmental history studies', *Frontiers of History in China*, 2, pp. 121–44.

Mellanby, K. (1981), *Farming and Wildlife*, London.

Michie, J. G. (ed.) (1901), *Records of Invercauld*, New Spalding Club, Aberdeen.

Mizoguchi, T. (ed.) (2008), *The Environmental Histories of Europe and Japan*, Nagoya University.

Moore, N. (1987), *The Bird of Time: the Science and Politics of Nature Conservation*, Cambridge.

Moseley, S. (2001), *The Chimney of the World: a History of Smoke Pollution in Victorian and Edwardian Manchester*, Cambridge.

Nairne, D. (1982), 'Notes on Highland woods, ancient and modern', *Transactions of the Gaelic Society of Inverness*, 17, pp. 170–221.

Nicholson, M. (1951), *Birds and Men*, London.

Olwig, K. (1984), *Nature's Ideological Landscape*, London.

Peterken, G. F. (1996), *Natural Woodland: Ecology and Conservation in Northern Temperate Regions*, Cambridge.

Rackham, O. (1986), *History of the Countryside*, London.

Rackham, O. (2003), *Ancient Woodland, its History, Vegetation and Uses in England*, London.

Rackham, O. (2006), *Woodlands*, London.

Ritchie, J. (1920), *Influence of Man on Animal Life in Scotland*, Cambridge.

Rodger, D., Stokes, J. and Ogilvie, J. (2006), *Heritage Trees of Scotland*, Edinburgh.

Sansum, P. (2004), 'Historical resource use and ecological change in semi-natural woodland: western oakwoods in Argyll, Scotland', unpublished University of Stirling Ph.D. thesis.

Sheail, J. (1975), 'The concept of national parks in Great Britain, 1900–1950', *Transactions of the Institute of British Geographers*, 65, pp. 41–56.

Sheail, J. (1976), *Nature in Trust: the History of Nature Conservation in Britain*, Glasgow.

Sheail, J. (1981), *Rural Conservation in Inter-war Britain*, Oxford.

Sheail, J. (1985), *Pesticides and Nature Conservation: the British Experience 1950–1975*, Oxford.

Sheail, J. (1987), *Seventy-five Years in Ecology: the British Ecological Society*, Oxford.

Sheail, J. (2002), *An Environmental History of Twentieth-century Britain*, Basingstoke.

Short, B., Watkins, C. and Martin, J. (eds) (2006), *The Front Line of Freedom: British Farming in the Second World War*, British Agricultural History Society.

Shrubb, M. (2003), *Birds, Scythes and Combines: a History of Birds and Agricultural Change*, Cambridge.

Simmons, I. G. (2001), *An Environmental History of Great Britain from 10,000 Years Ago to the Present*, Edinburgh.

Simmons, I. G. (2003), *The Moorlands of England and Wales: an Environmental History 8000 BC–2000*, Edinburgh.

Simmons, I. and Tooley, M. (eds) (1981), *The Environment in British Prehistory*, Cornell.

Slack, P. (ed.) (1999), *Environments and Historical Change: the Linacre Lectures 1998*, Oxford.

Smith, A. G. (1971), 'The influence of Mesolithic and Neolithic man on British vegetation: a discussion', in D. Walker and R. G. West (eds), *Studies in the Vegetational History of the British Isles*, Cambridge, pp. 81–96.

Smout, T. C. (ed.) (1993), *Scotland since Prehistory: Natural Change and Human Impact*, Aberdeen.

Smout, T. C. (ed.) (1997), *Scottish Woodland History*, Edinburgh.

Smout, T. C. (2000), *Nature Contested: Environmental History in Scotland and Northern England since 1600*, Edinburgh.

Smout, T. C. (ed.) (2001), *Nature, Landscape and People since the Second World War*, Edinburgh.

Smout, T. C. (ed.) (2003), *People and Woods in Scotland: a History*, Edinburgh.

Smout, T. C. and Lambert, R. A. (eds) (1999), *Rothiemurchus: Nature and People on a Highland Estate, 1500-2000*, Edinburgh.

Smout, T. C., MacDonald, A. R., Watson, F. (2005), *A History of the Native Woodlands of Scotland, 1500–1920*, Edinburgh.

Sörlin, S. and Warde, P. (2007), 'The problem of the problem of environmental history: a re-reading of the field', *Environmental History*, 12, pp. 107–30.

Steven, H. M. and Carlisle, A. (1959), *The Native Pinewoods of Scotland*, Edinburgh.

Stevenson, G. B. (2007), 'An historical account of the social and economic causes of capercaillie *Tetrao urogallus* extinction and reintroduction in Scotland', unpublished University of Stirling Ph.D. thesis.

Tachibana, S. and Watkins, C. (2008), 'The assimilation, naturalisation and hybridisation of Japanese trees and shrubs in Britain 1700–1920', in T. Mizoguchi (ed.), *The Environmental Histories of Europe and Japan*, Nagoya, pp. 185–192.

Thomas, K. (1983), *Man and the natural World: Changing Attitudes in England 1500–1800*, Harmondsworth.

Thompson, D. B. A., Hester, A. J. and Usher, M. B (1995), *Heaths and Moorlands: Cultural Landscapes*, Edinburgh.

Tipping, R. (1994), 'The form and fate of Scottish woodlands', *Proceedings of the Society of Antiquaries of Scotland*, 124, pp. 1–54.

Tipping, R. (2003), 'Living in the past: woods and people in prehistory to 1000 BC', in T. C. Smout (ed.), *People and Woods in Scotland*, Edinburgh, pp. 14–39.

Tittensor, R. M. (1970), 'History of the Loch Lomond oakwoods', *Scottish Forestry*, 24 (1970), pp. 100–18.

Vera, F. W. M. (2000), *Grazing Ecology and Forest History*, Wallingford.

Warde, P. (2008), *Energy Consumption in England and Wales, 1560–2000*, Naples.

Watkins, C. (1998), *European Woods and Forests, Studies in Cultural History*, Wallingford.

Watson, A. (1983), 'Eighteenth century deer numbers and pine regeneration near Braemar, Scotland', *Biological Conservation*, 24, pp. 289–305.

Whitehouse, A. J. (2004), 'Negotiating small differences: conservation organisations and farming in Islay', unpublished University of St Andrews Ph.D. thesis.

Williams, R. (1973), *The Country and the City*, London.

Winchester, A. J. L. (2000), *The Harvest of the Hills: Rural Life in Northern England and the Scottish Borders, 1400–1700*, Edinburgh.

Winter, J. (1999), *Secure from Rash Assault: Sustaining the Victorian Environment*, Berkeley.

Wormell, P. (2003), *Pinewoods of the Black Mount*, Skipton.

Worster, D. (ed.) (1988), *The Ends of the Earth: Perspectives on Modern Environmental History*, Cambridge.

Wrigley, E. A. (1988), *Continuity, Chance and Change: the Character of the Industrial Revolution in England*, Cambridge.

Wrigley, E. A. (1999), 'Meeting human energy needs: constraints, opportunities and effects', in P. Slack (ed.), *Environments and Historical Change: the Linacre Lectures 1998*, Oxford, pp. 76–95.

Wrigley, E. A. (2006), 'The transition to an advanced organic economy: half a millennium of English agriculture', *Economic History Review*, 59, pp. 435–80.

Yalden, D. (1999), *The History of British Mammals*, London.

Index

Page numbers in **bold** indicate an illustration.